CW01021258

He who calls what has vanished back again into being, enjoys a bliss like that of creating.

BARTHOLD NIEBUHR (1776–1831)

The real voyage of discovery consists not in seeking new landscapes, but in having new eyes.

MARCEL PROUST (1871–1922)

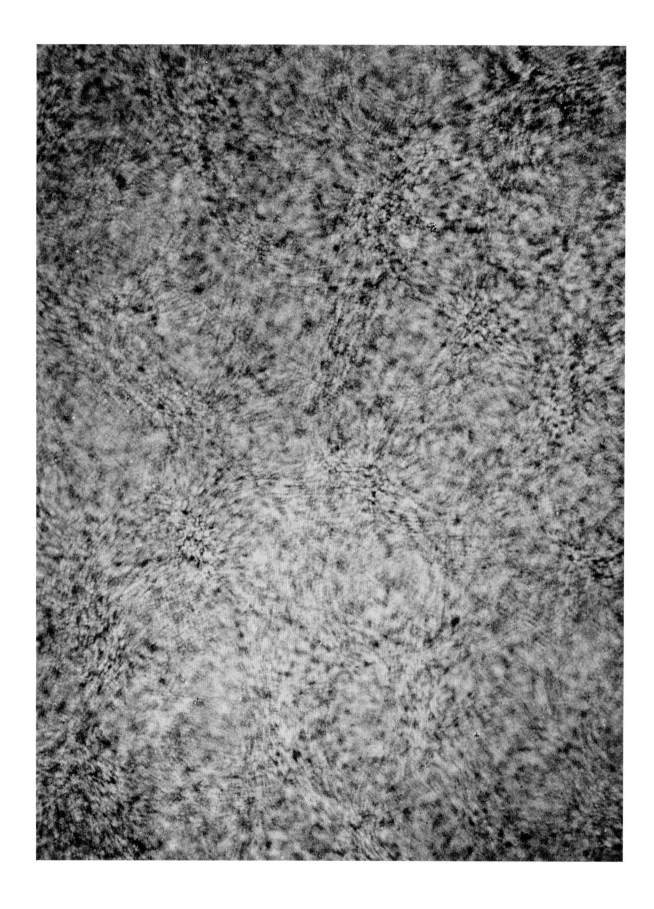

From *Annales de l'Observatoire*, S E E P.191

George Loudon

Object Lessons

The Visualisation of Nineteenth-Century Life Sciences

With photography by Rosamond Purcell

Lynne Cooke in conversation with George Loudon

Essay by Robert McCracken Peck

RIDINGHOUSE

Contents

For Angie, Emily and Marcus

Introduction

The material I have collected here seems to me, with some hindsight, to follow seamlessly on from 25 years of collecting contemporary art by young artists. This book is about this collection. Although the collection has no name, it does have clear, if wide, parameters. It represents a search for beauty, wonder, or magic in individual objects and illustrations related to nineteenth-century life sciences.

The nineteenth century witnessed the enormous breakthrough of science. The objects and illustrations in this collection express the wonder at the new that was felt at that time. They were made by skilled artists and craftsmen with longstanding traditions. Later on this material was made on an industrial scale for museums, schools and universities, generally no longer by artisans, hence less beautiful.

The material I collect has lost its original purpose. It has disappeared from view in museums and universities and been consigned to storage. But by losing its original purpose it has become open to new meanings and especially new visual interpretations. This is what I believe has appealed to contemporary artists who have either appropriated or been inspired by these visuals.

My conversation with Lynne Cooke and the essay by Robert McCracken Peck provide more background to these considerations and separate the two distinct sections into which the book is divided. The first part of the book shows some examples of what my 'eye' feels to have strong aesthetic appeal. This is of course a very subjective choice, but perhaps one with which some readers might agree. It is an invitation to look more closely. In the second part of the book, each individual work is presented and described in almost the order in which it was acquired.

I hope this book stimulates readers' curiosity and their appreciation of a new kind of beauty.

GEORGE LOUDON
July 2015

Face-reading model, SEE P.234

OVERLEAF From *Kai-Senshu*, SEE P.94

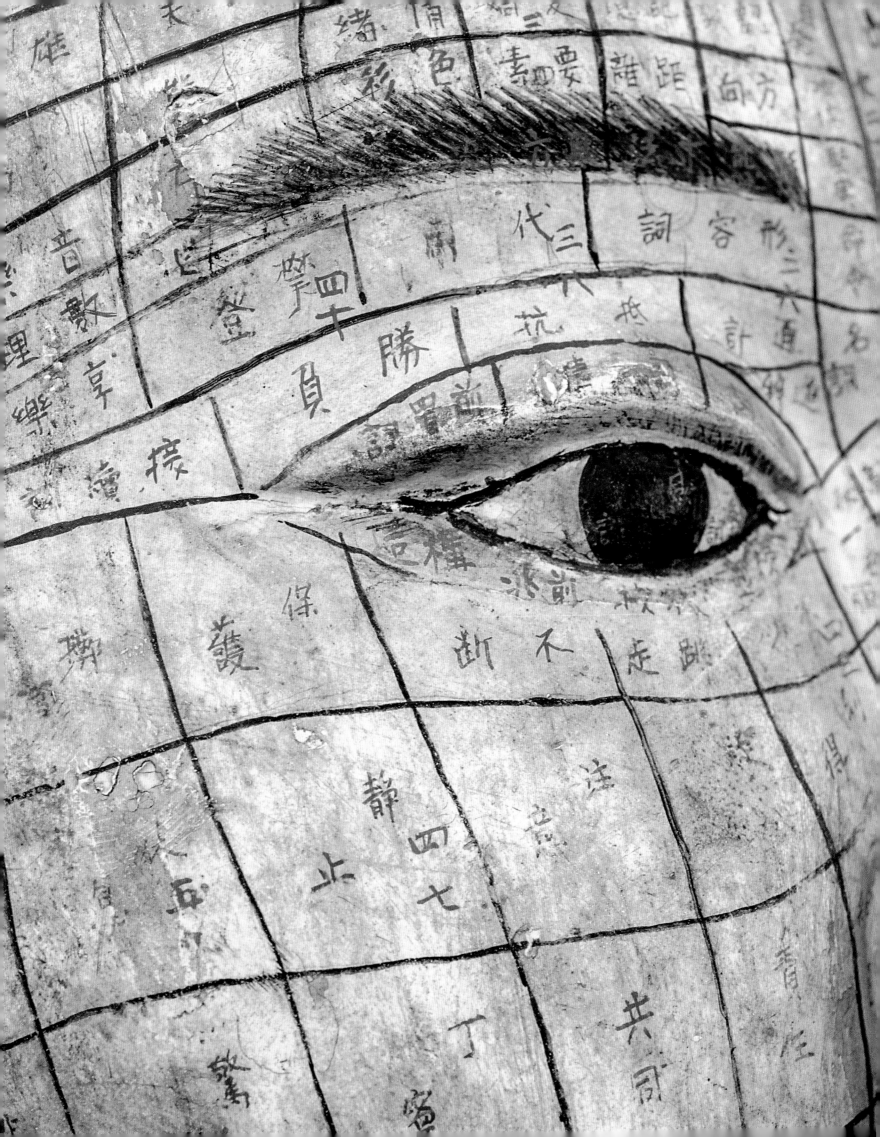

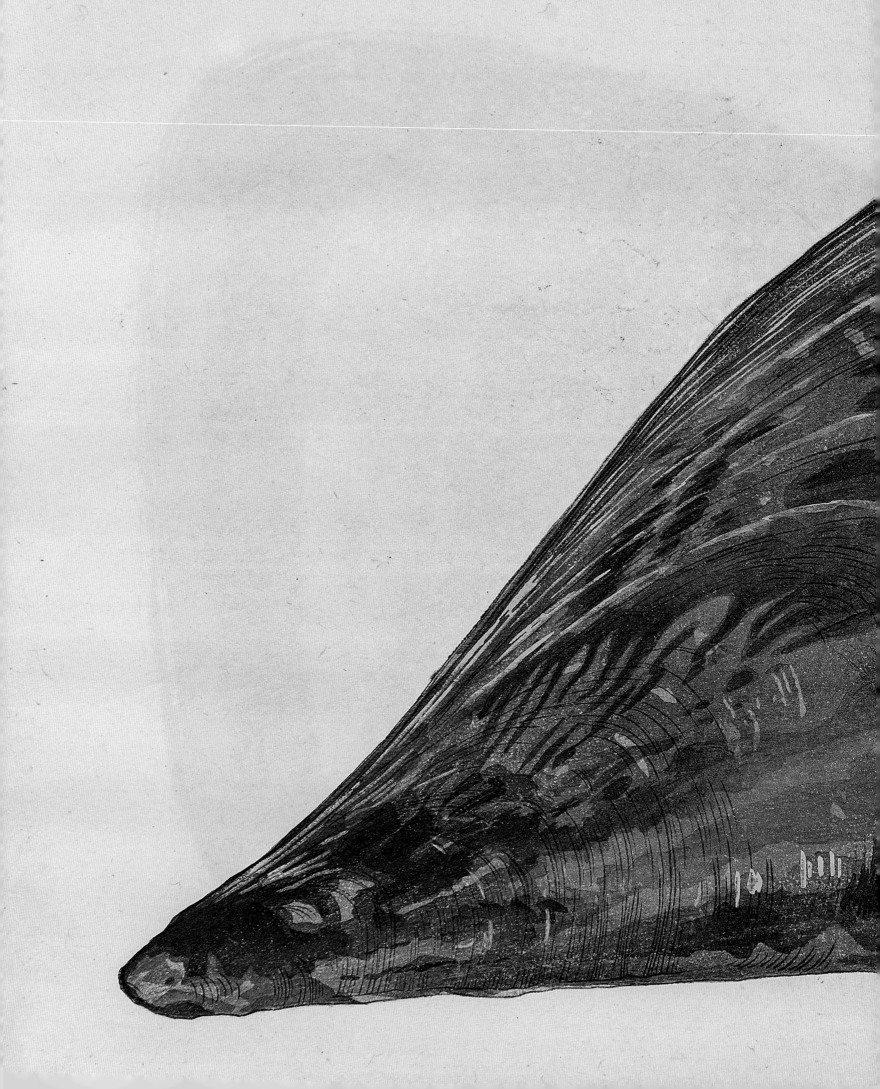

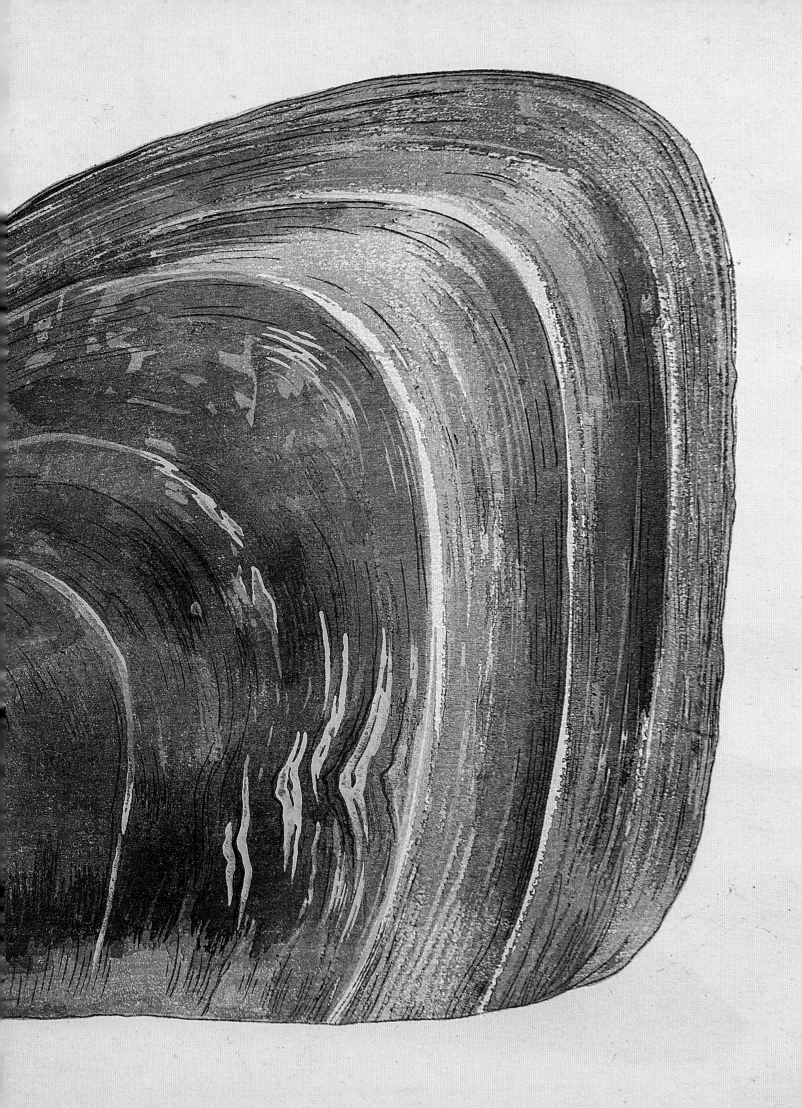

From *Wandtafeln für den Unterricht in der Naturkunde,* SEE P.163

OVERLEAF Medical heads, SEE P.89

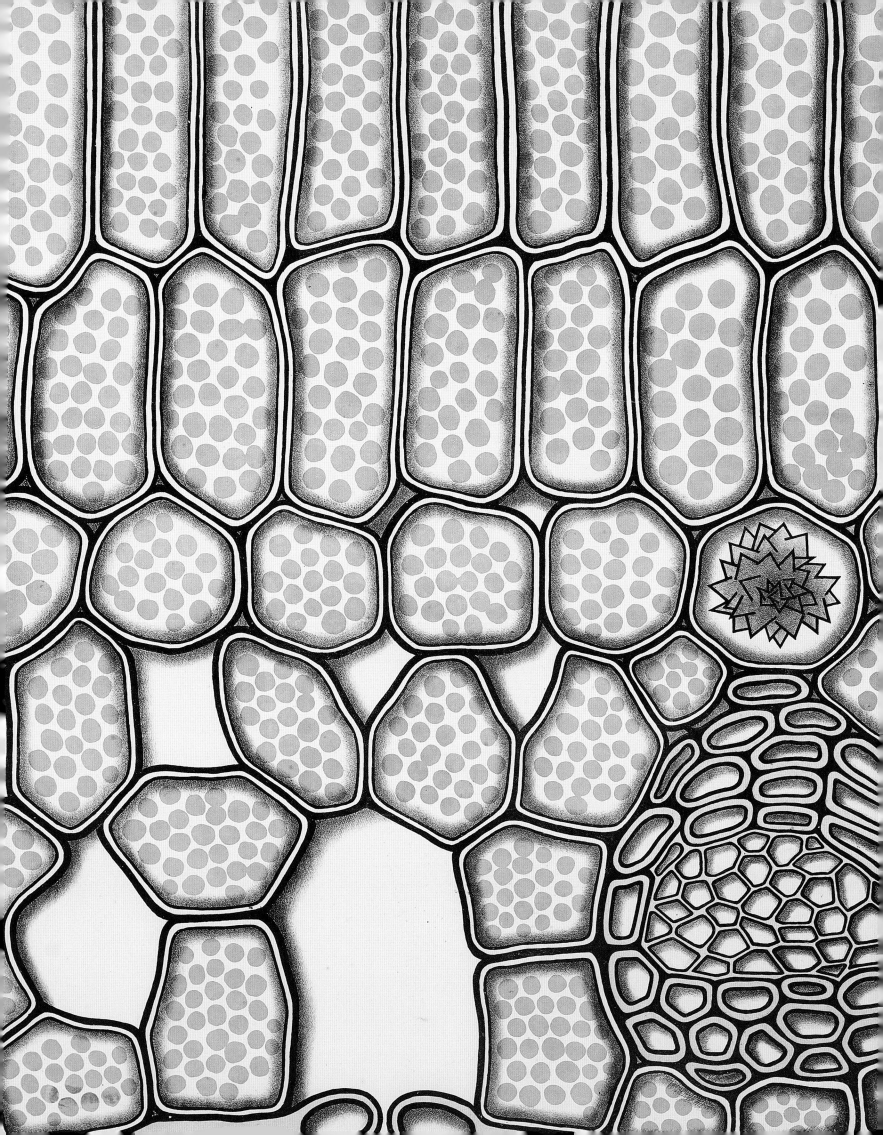

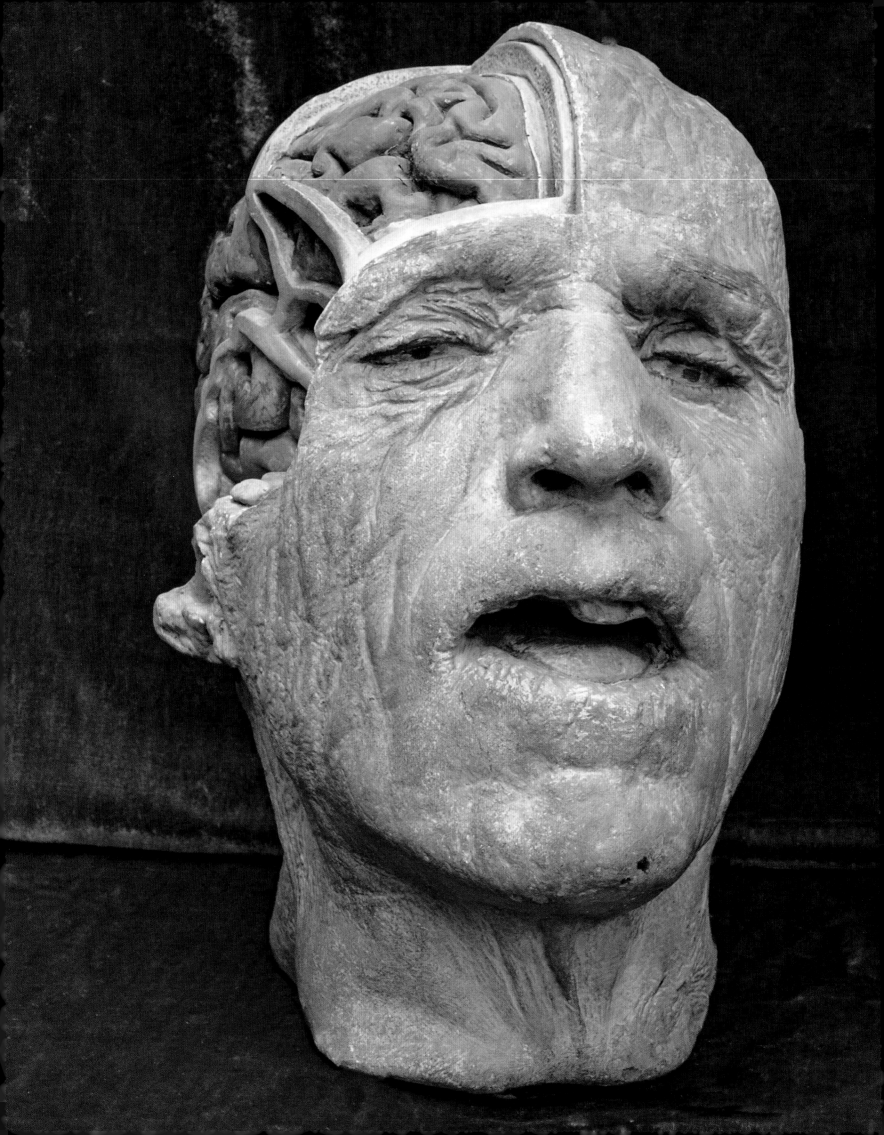

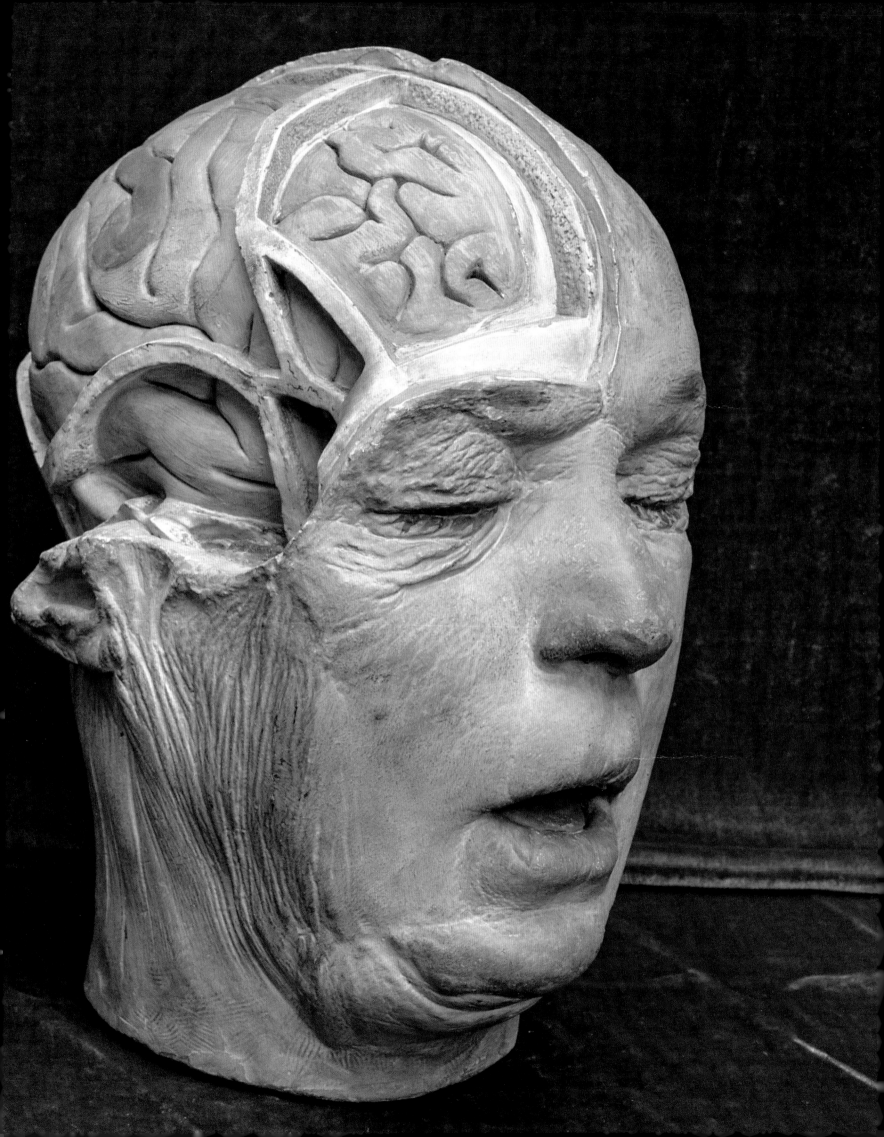

Elephant bird egg, SEE P.88

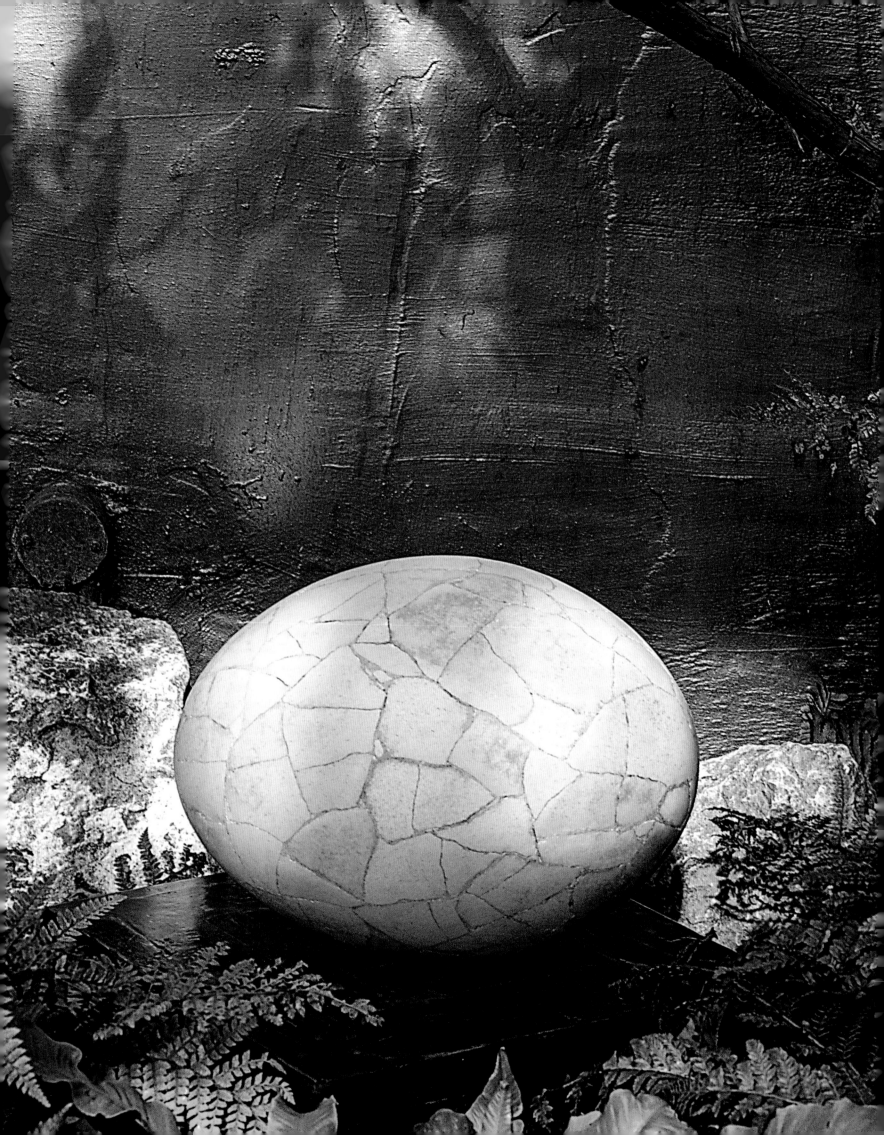

Album of ferns, SEE P.125

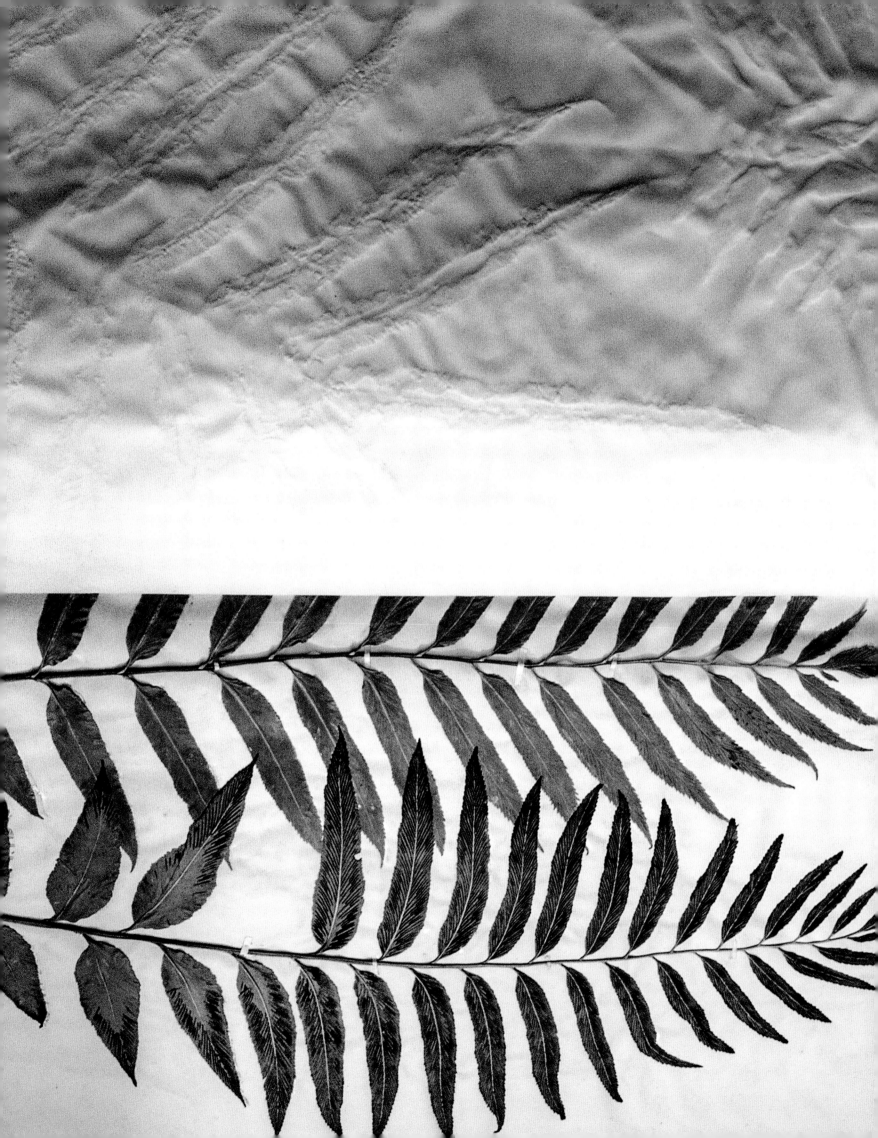

Exploded skull, SEE P.91

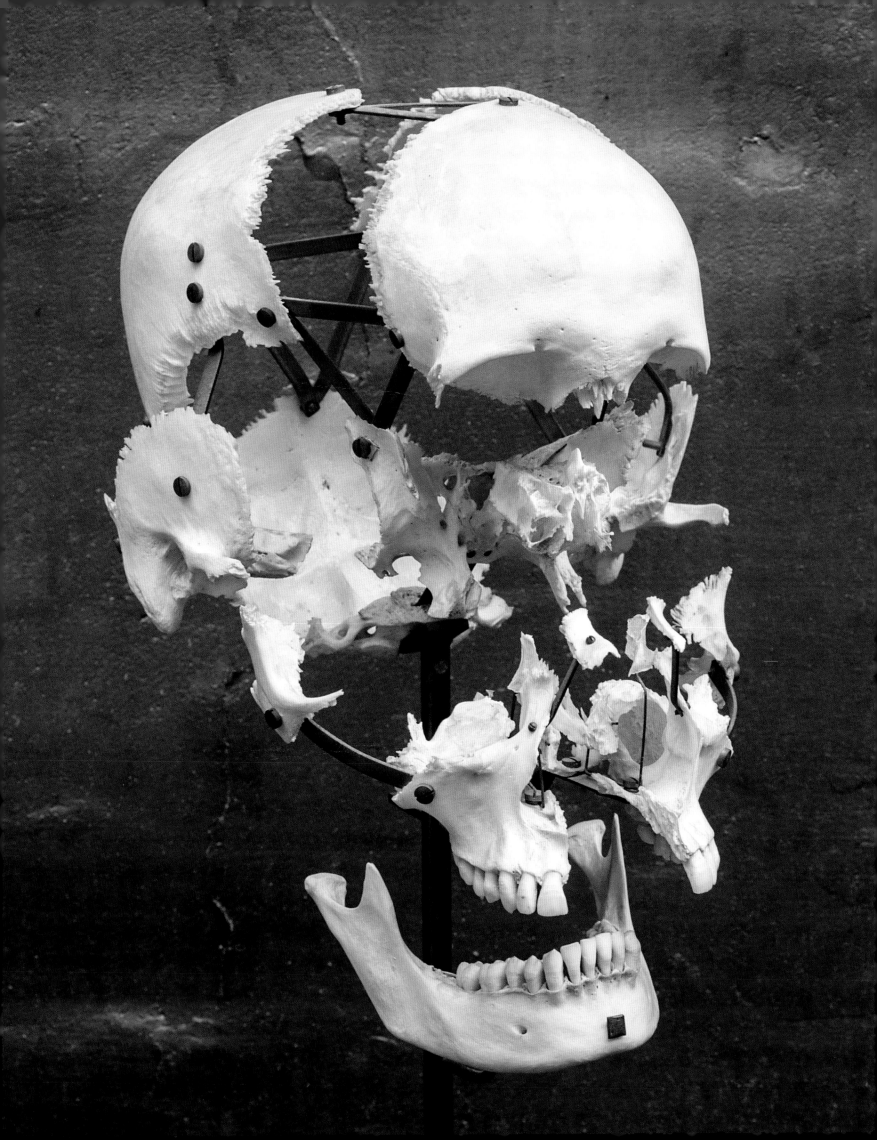

Model fruit and vegetables, SEE P.188
OVERLEAF Bottled fruit and vegetables, SEE P.165

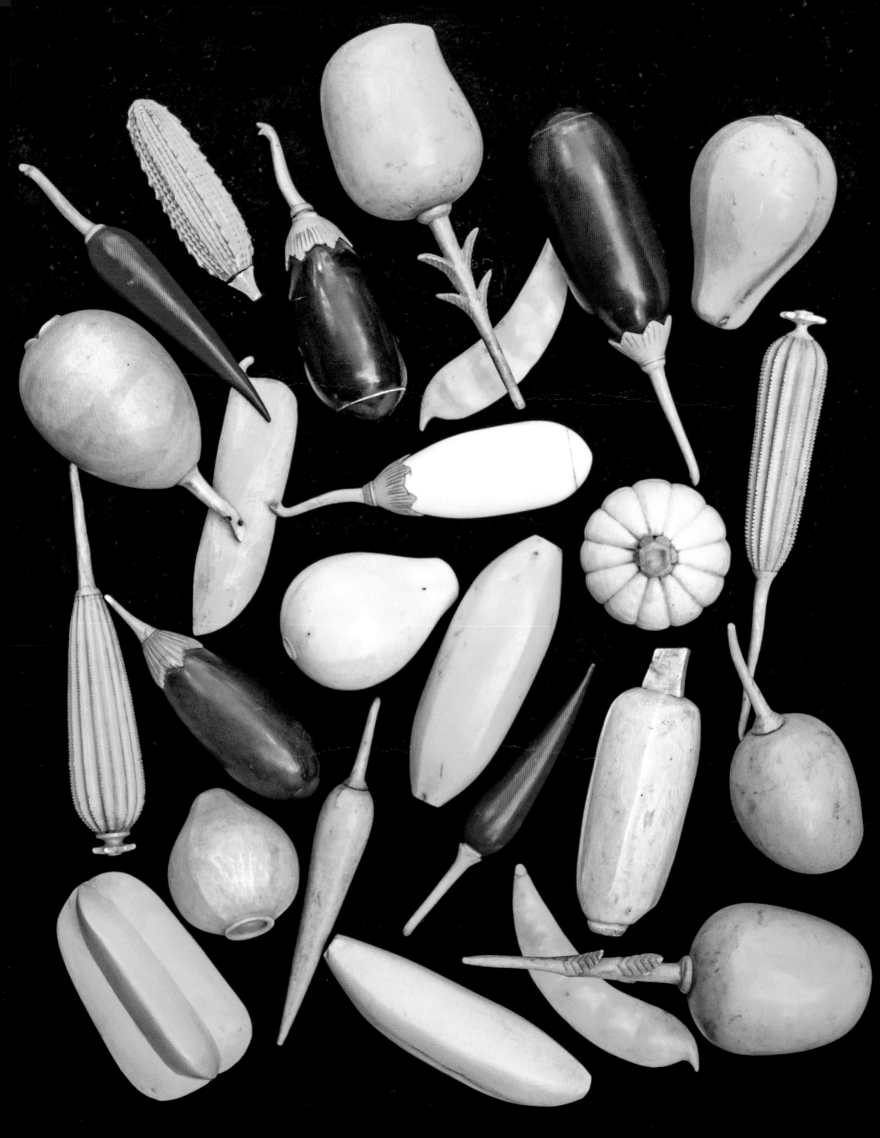

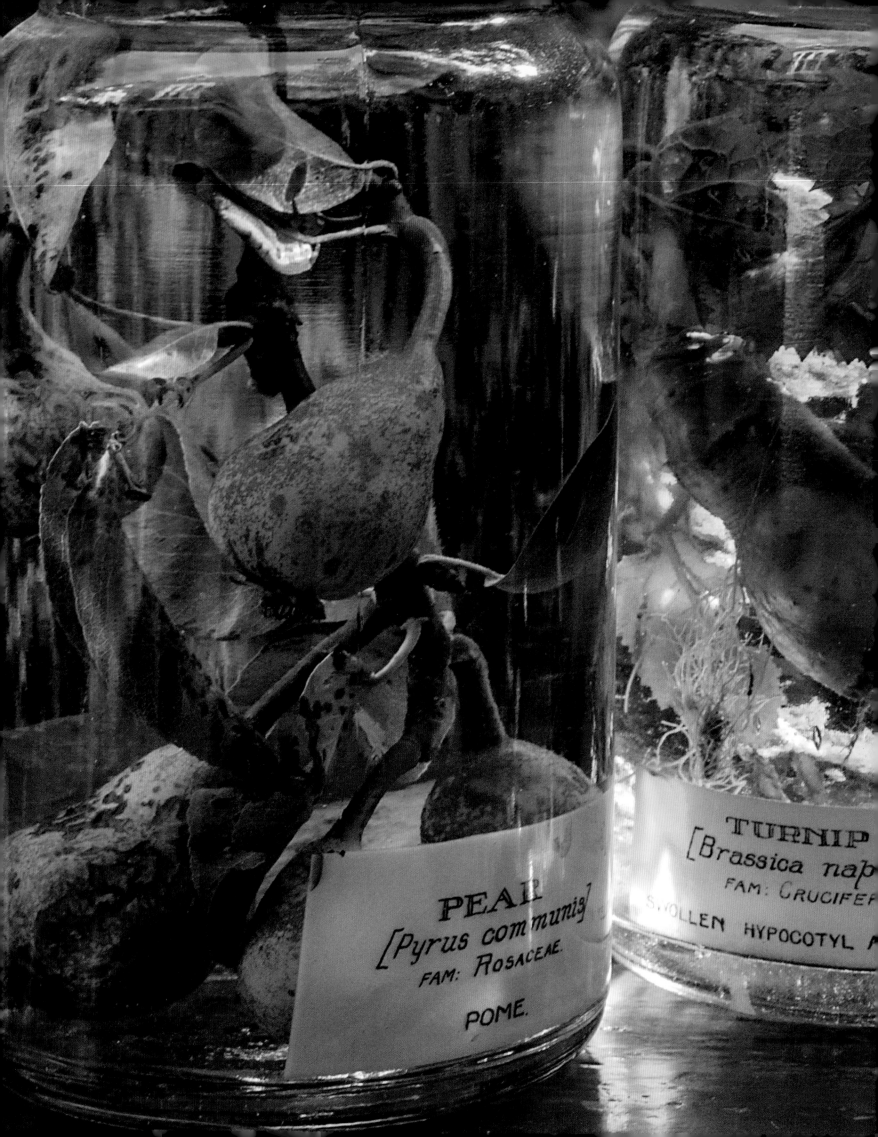

PEAR
[Pyrus communis]
FAM: Rosaceae.

POME.

TURNIP
[Brassica nap
FAM: Crucifer

SWOLLEN HYPOCOTYL

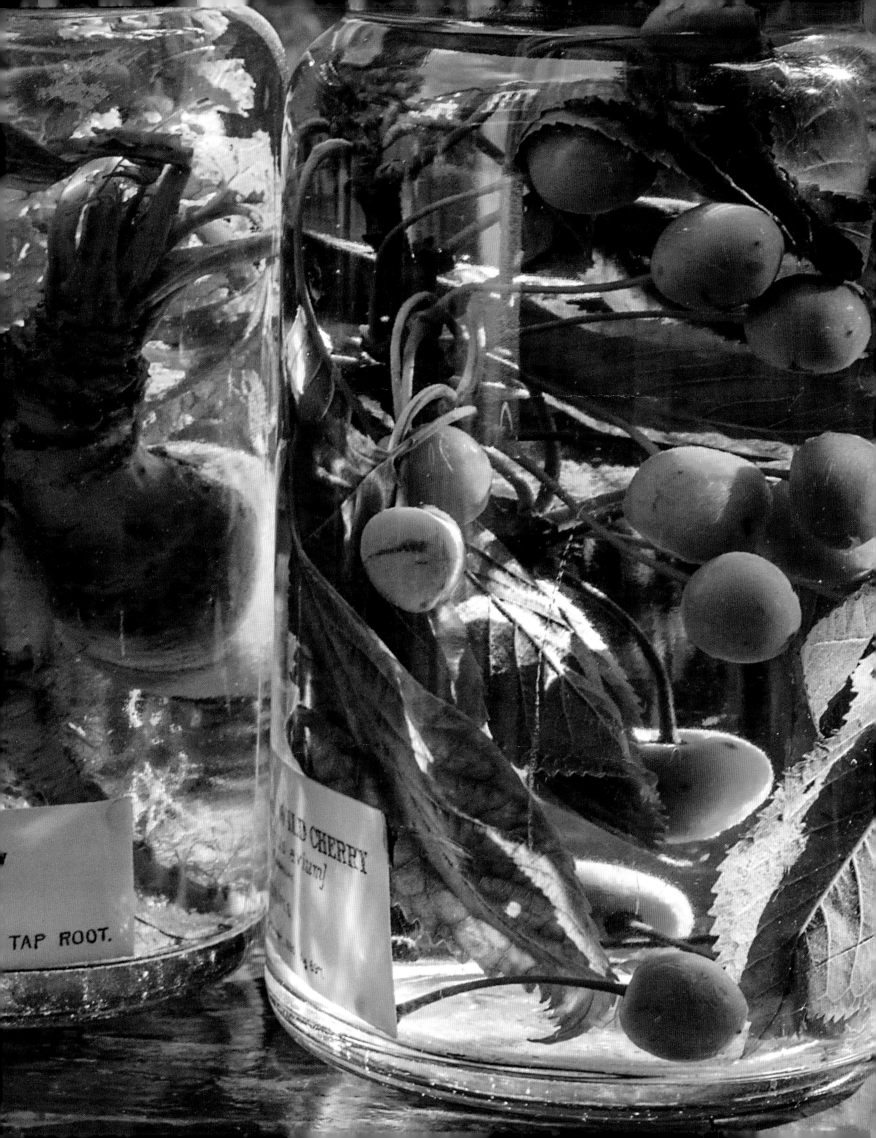

TAP ROOT.

WILD CHERRY

Model of squid, SEE P.194

OVERLEAF From *Fauna Japonica*, SEE P.119

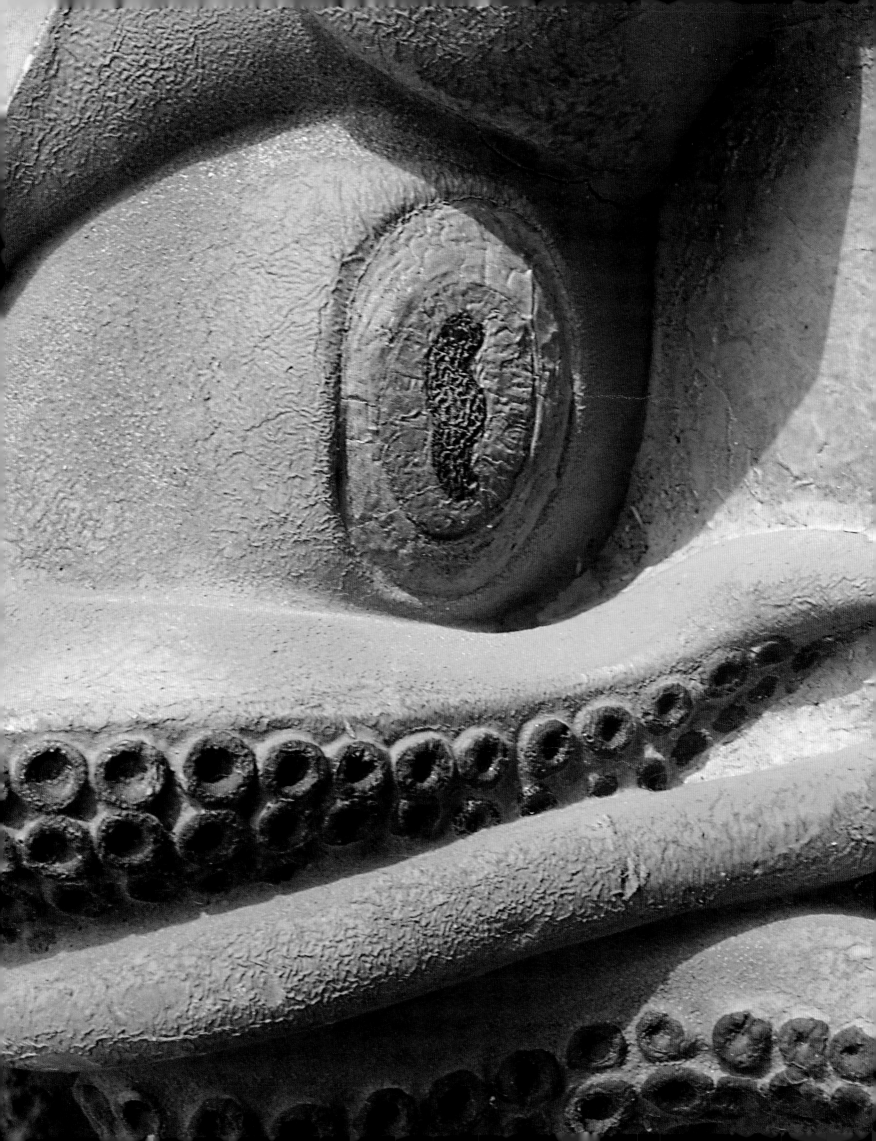

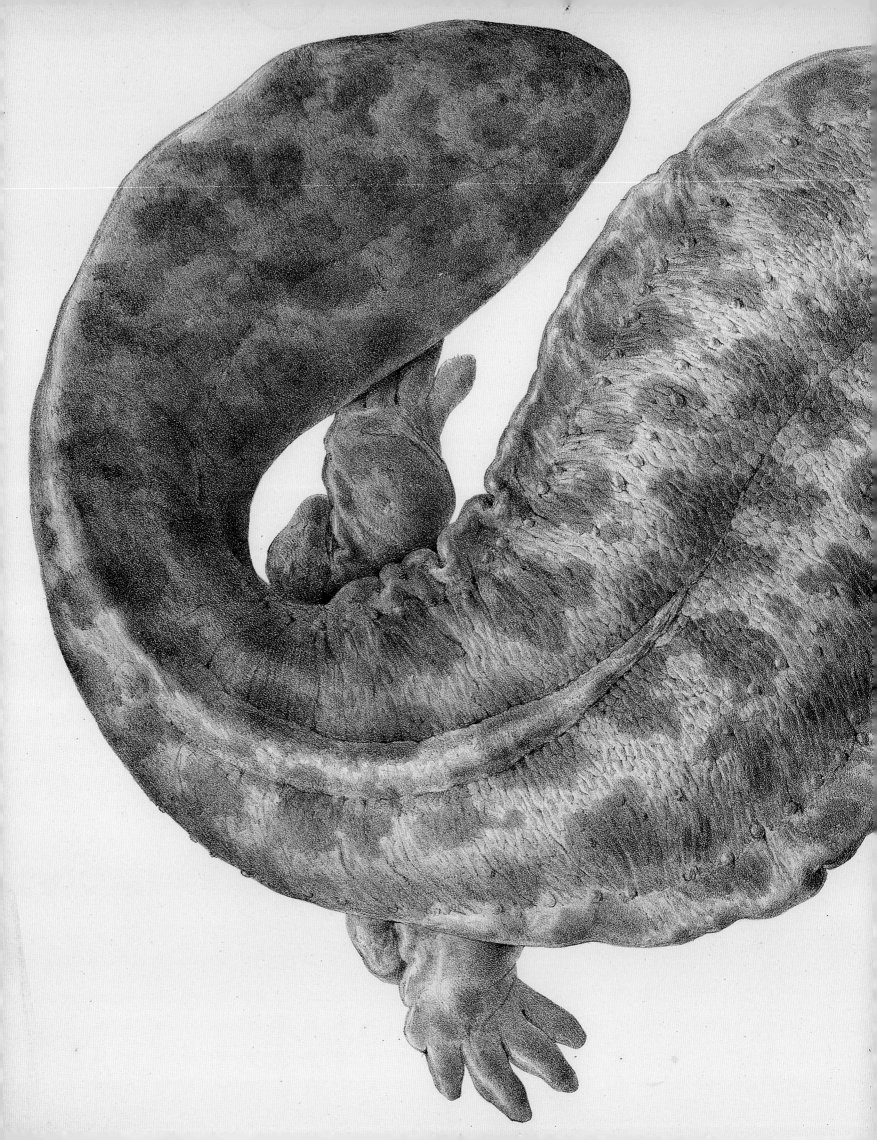

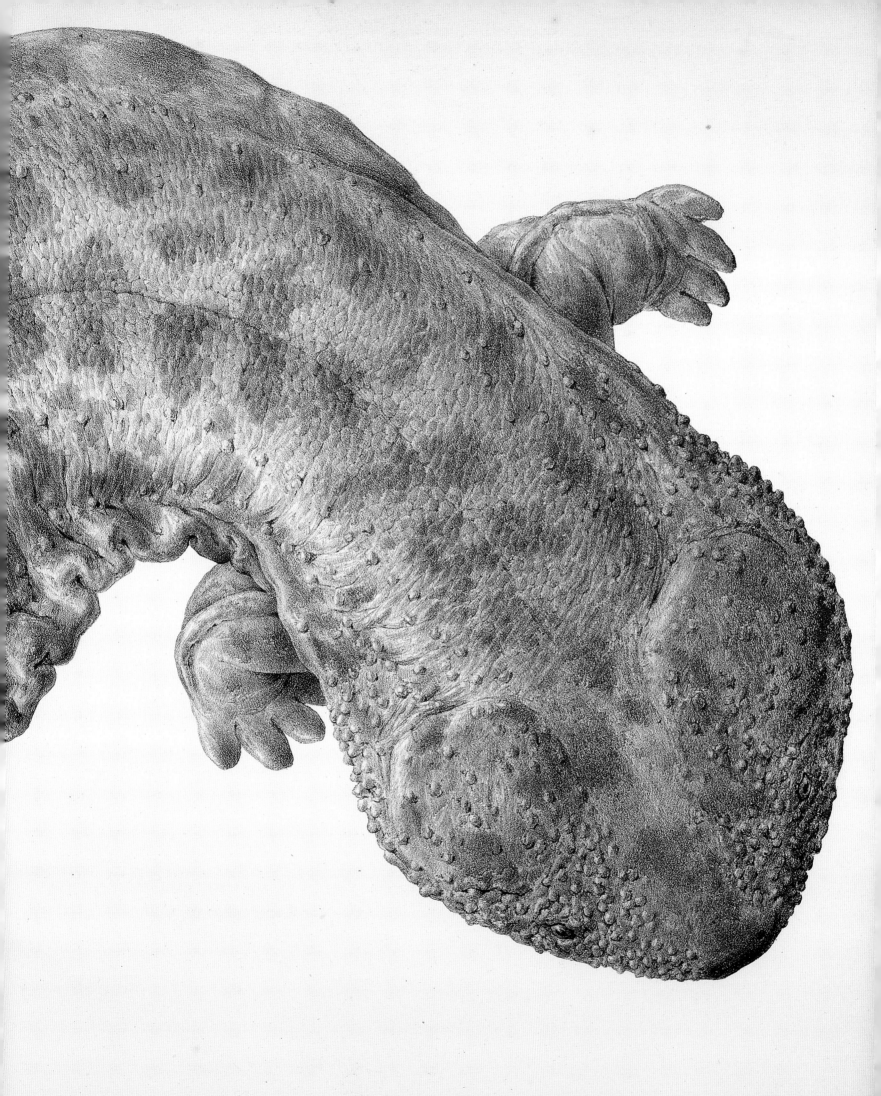

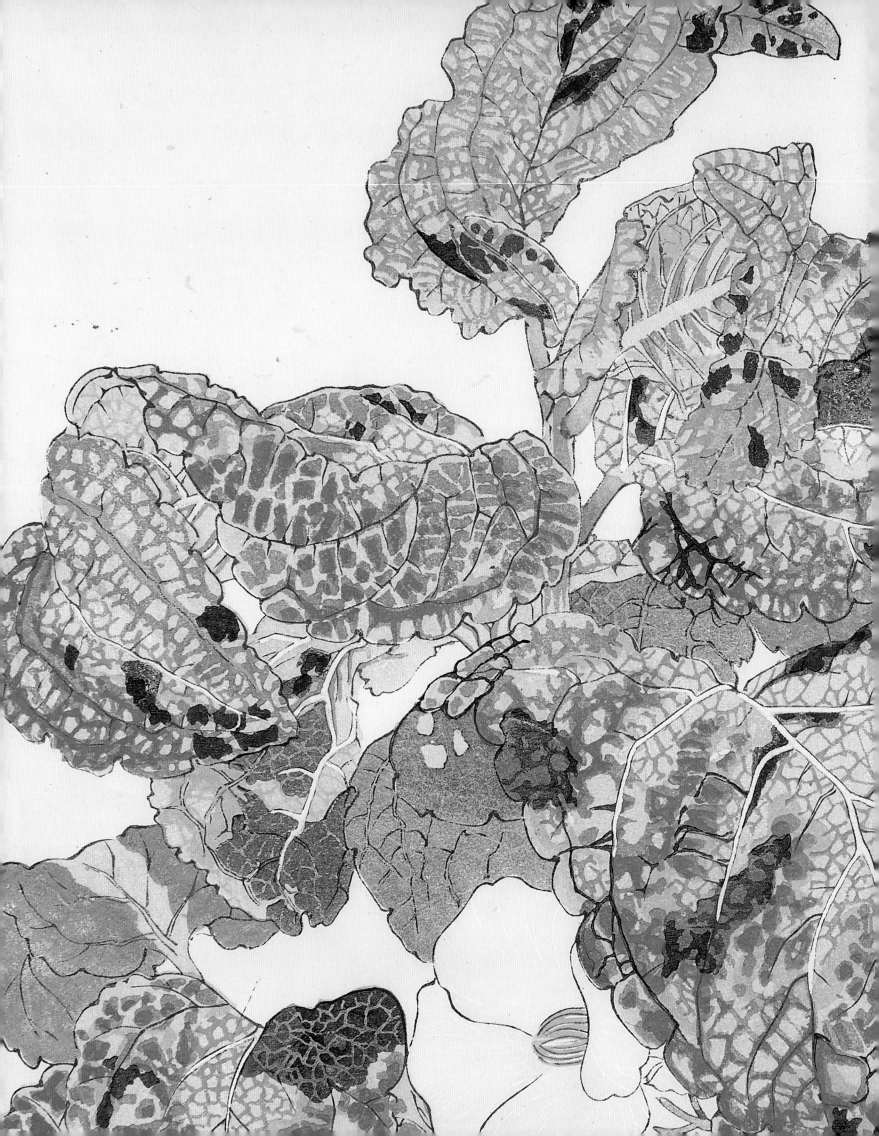

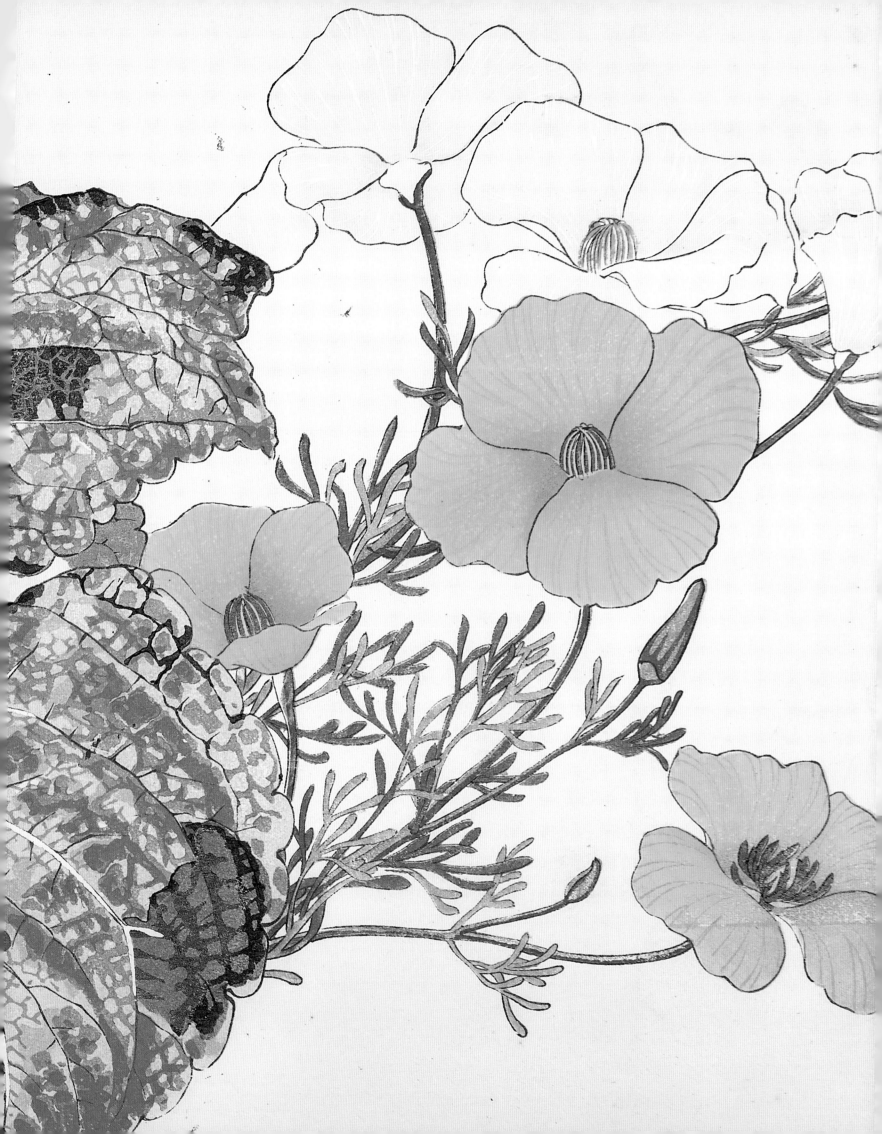

PREVIOUS From *Seiyō Kusabana Zufu*, SEE P.120

Hand axe, SEE P.103

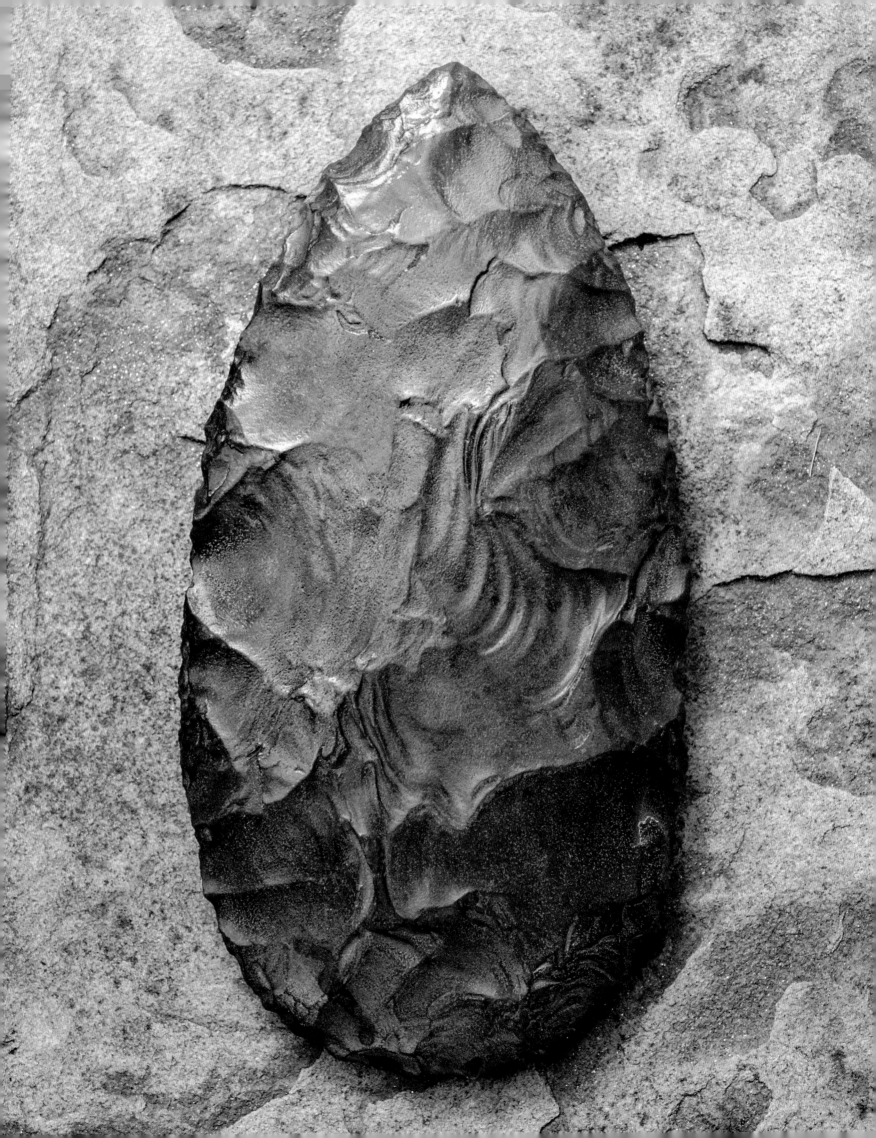

From *The Bakerian lecture on the total solar eclipse*, SEE P.172
OVERLEAF Models of mushrooms, SEE PP.170, 231

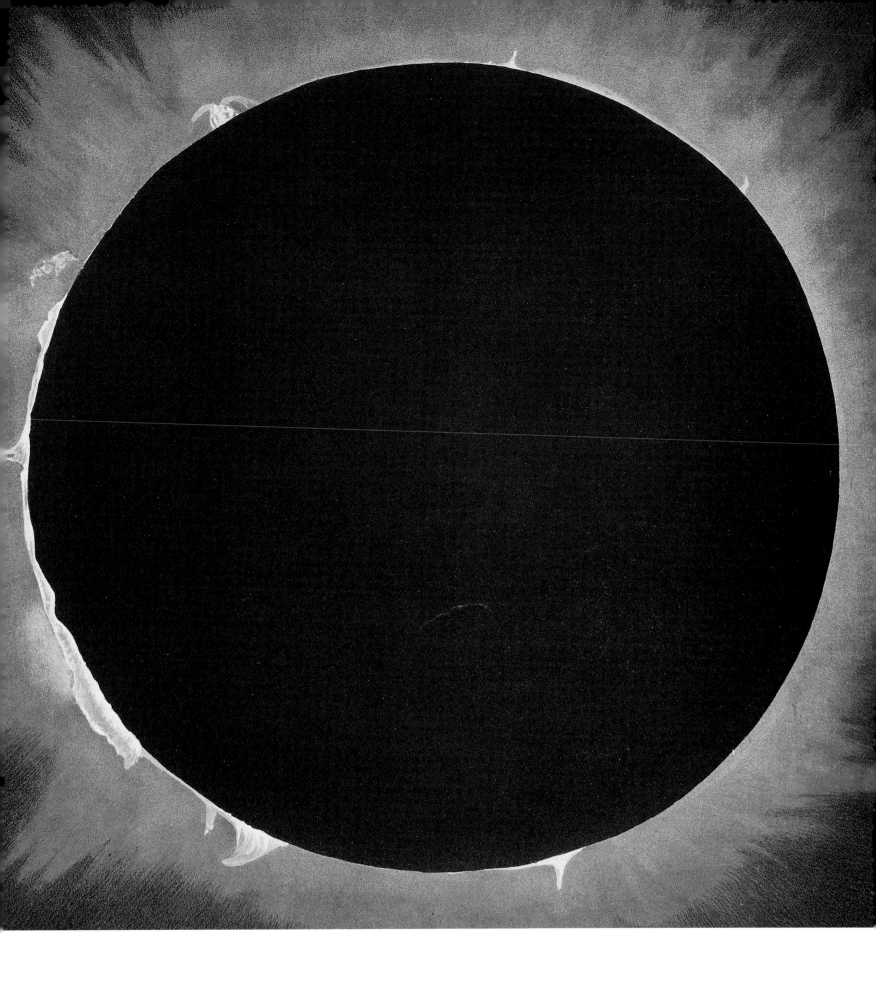

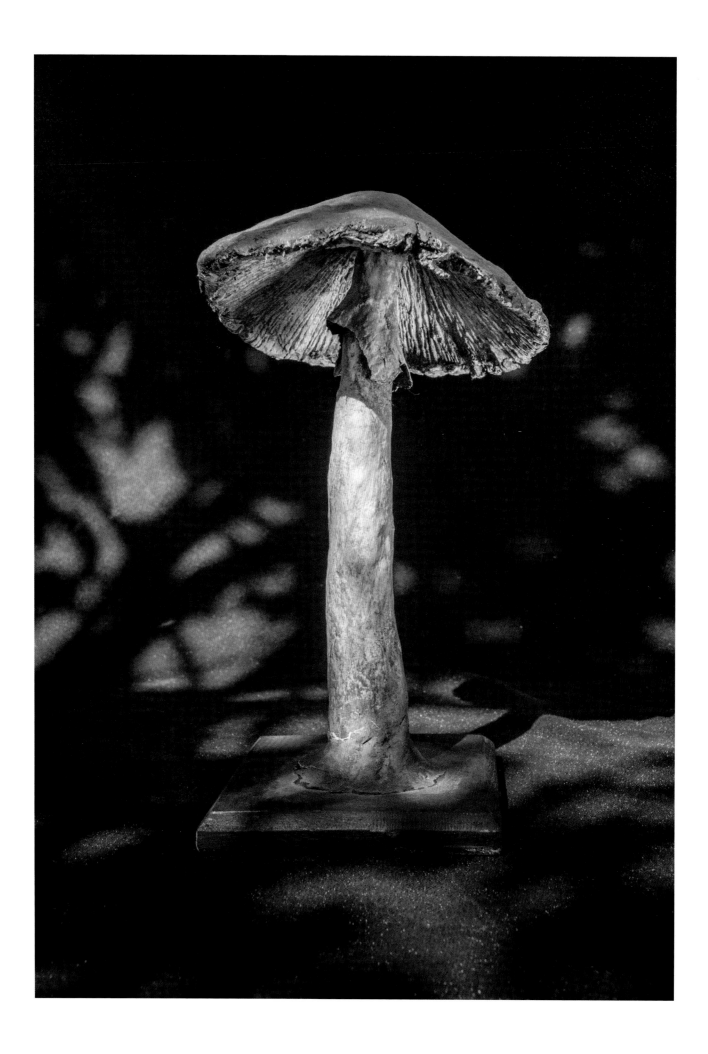

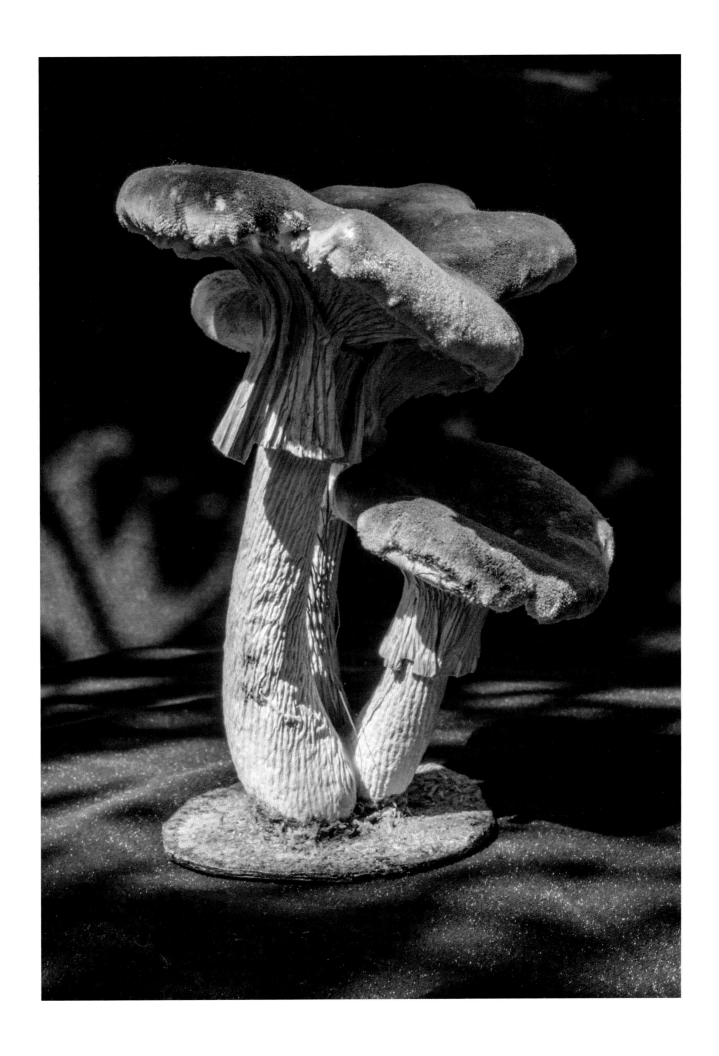

Conjoined piglets, SEE P.133
OVERLEAF Models of carnations, SEE PP.208–09

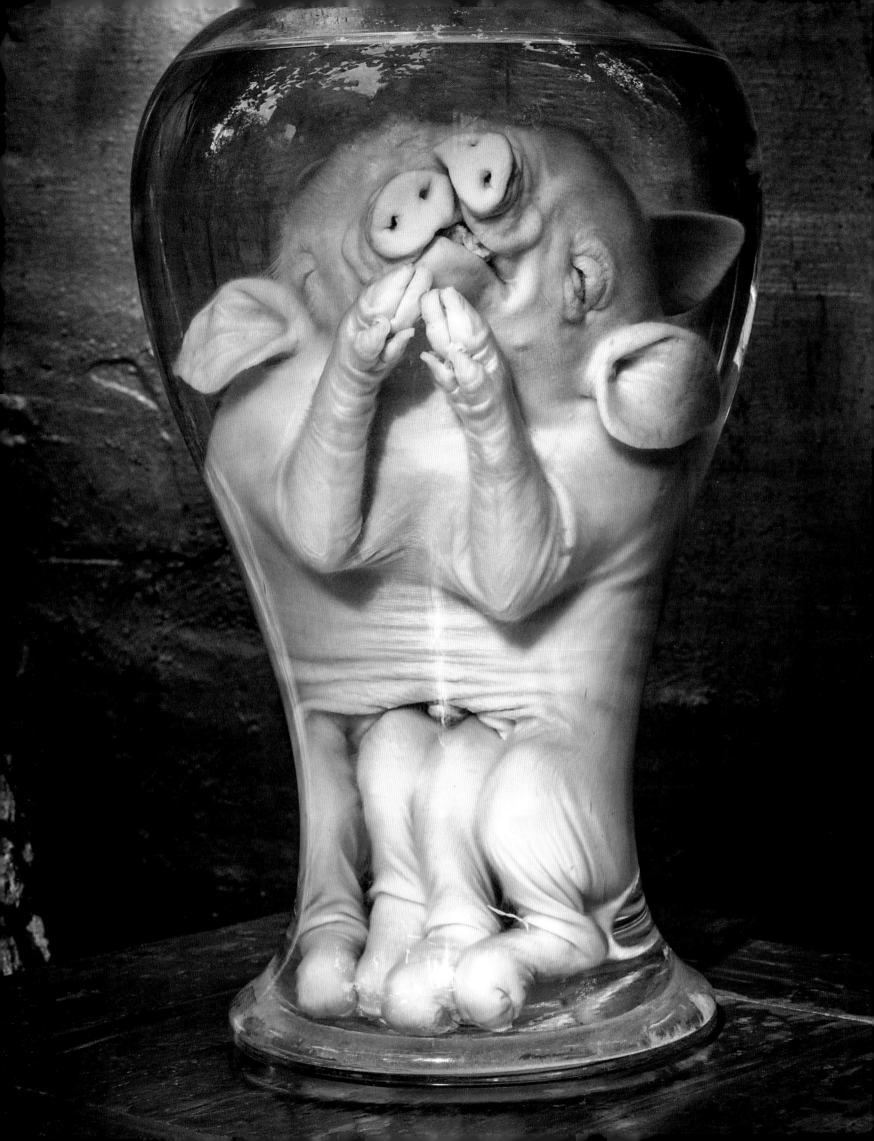

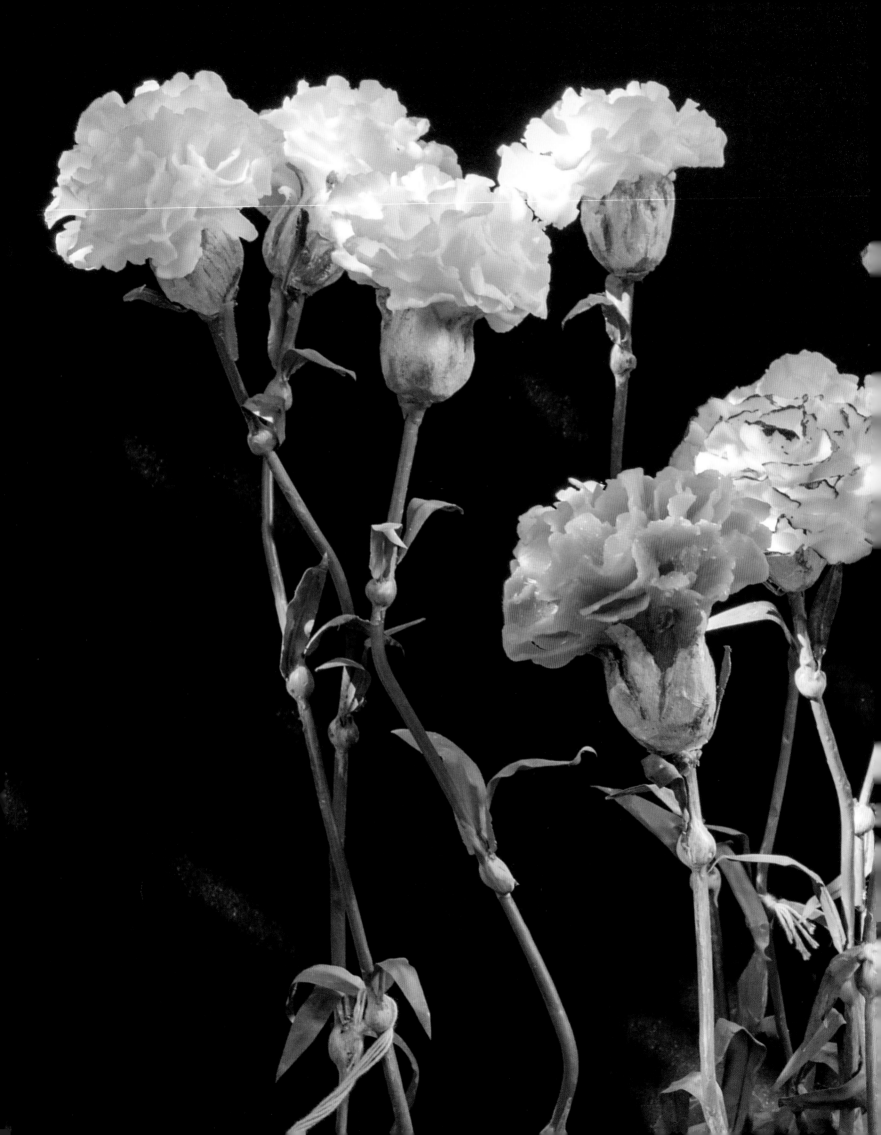

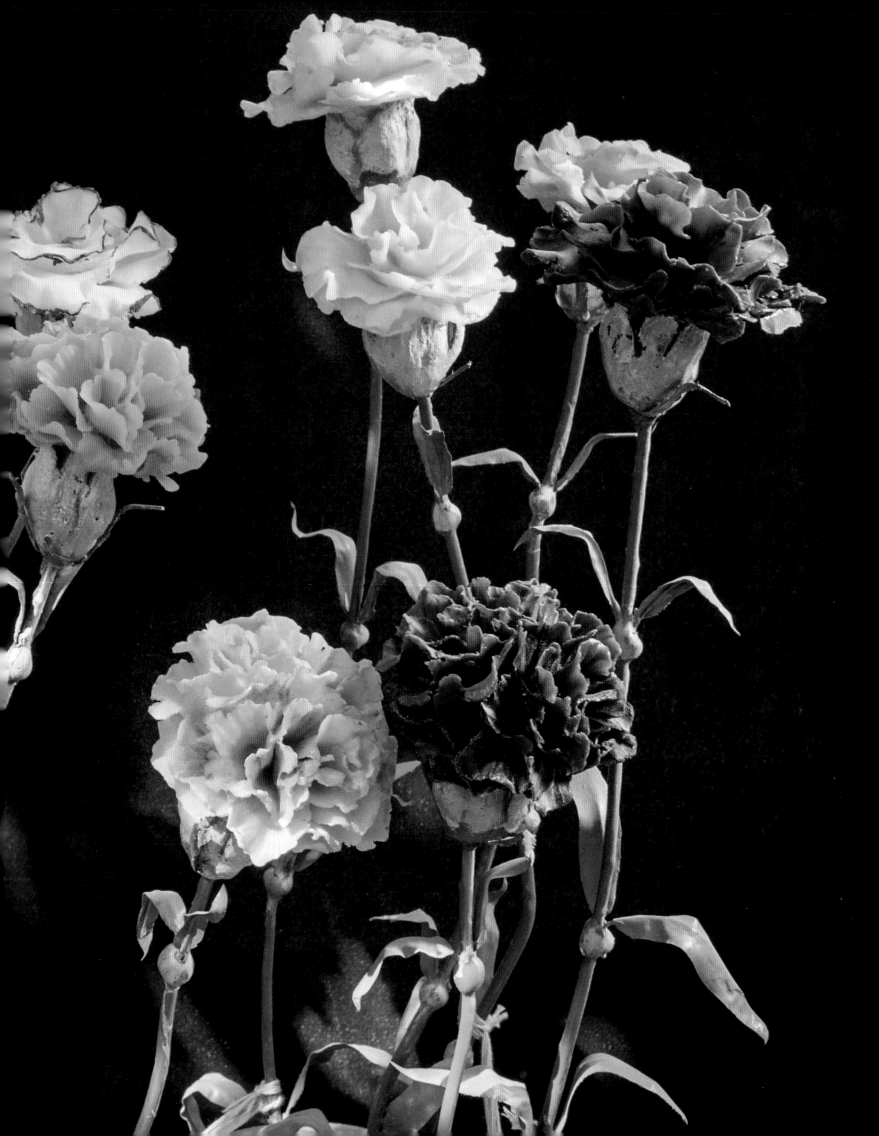

From *Mikroskopische Pflanzenbilder*, SEE P.167

OVERLEAF Botanical models, SEE PP.104–05

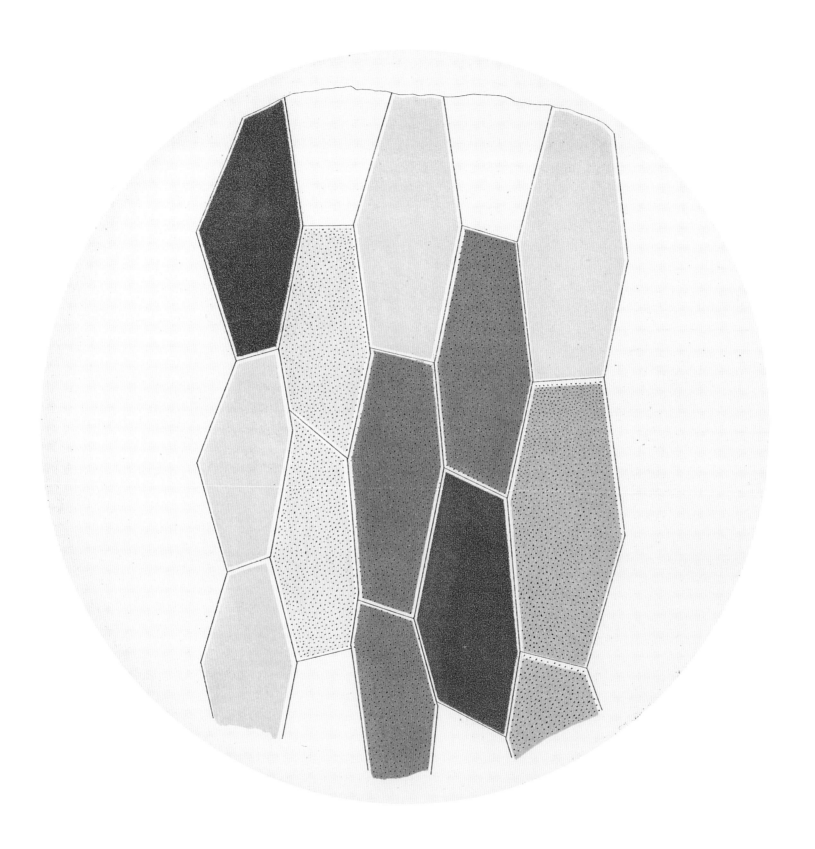

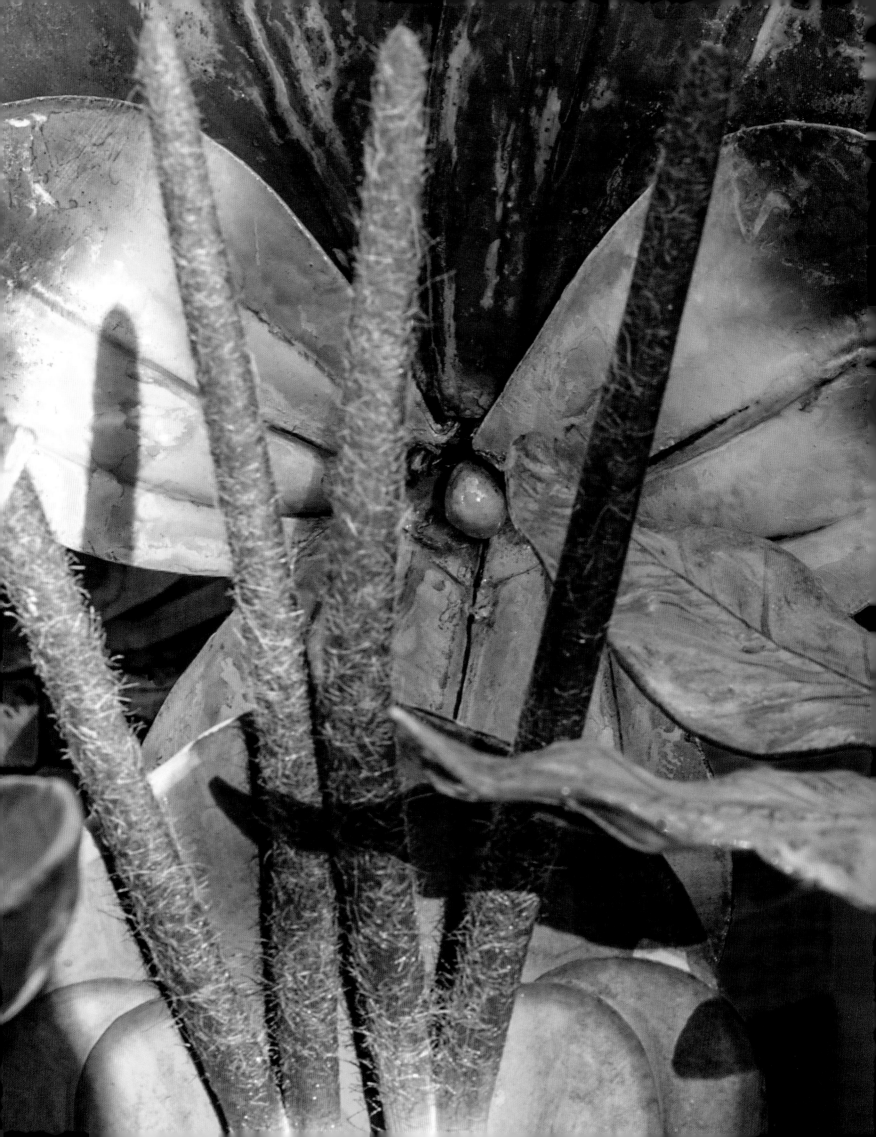

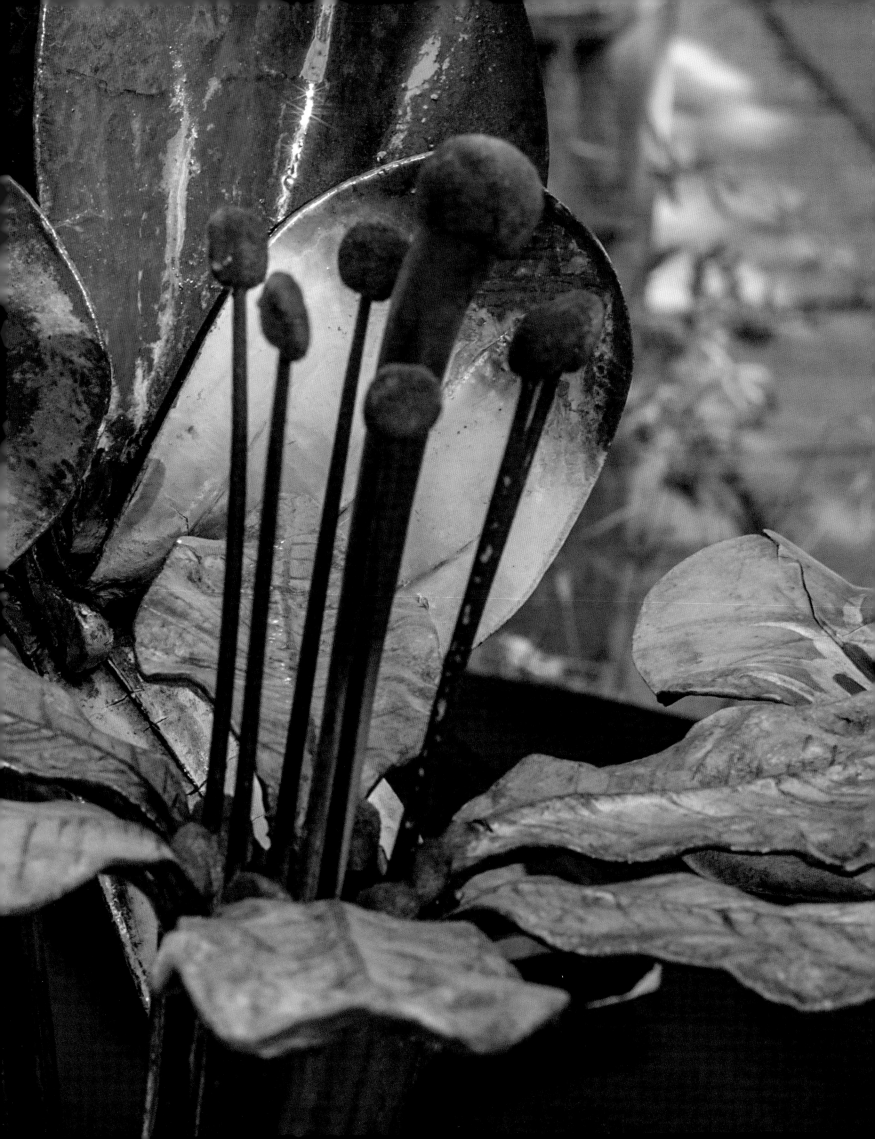

Two-faced kitten, SEE P.147
OVERLEAF Models of mushrooms, SEE P.170

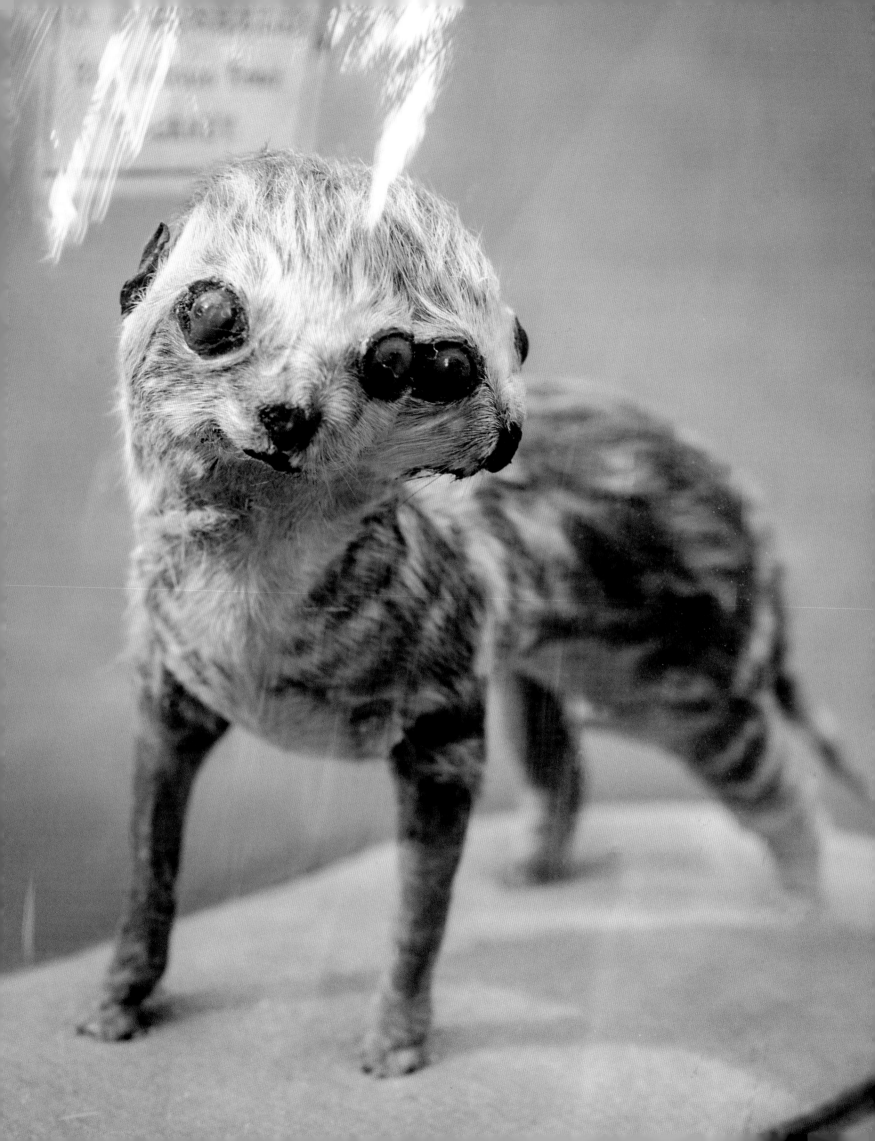

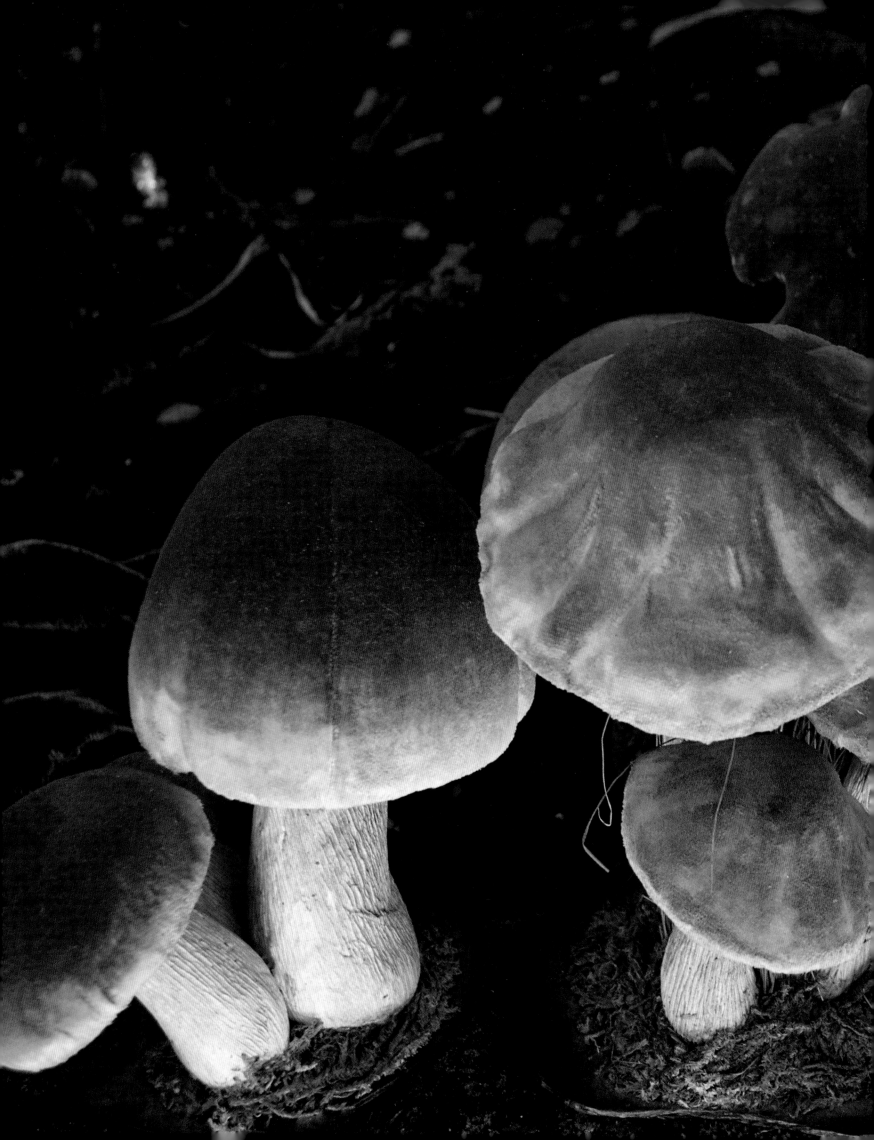

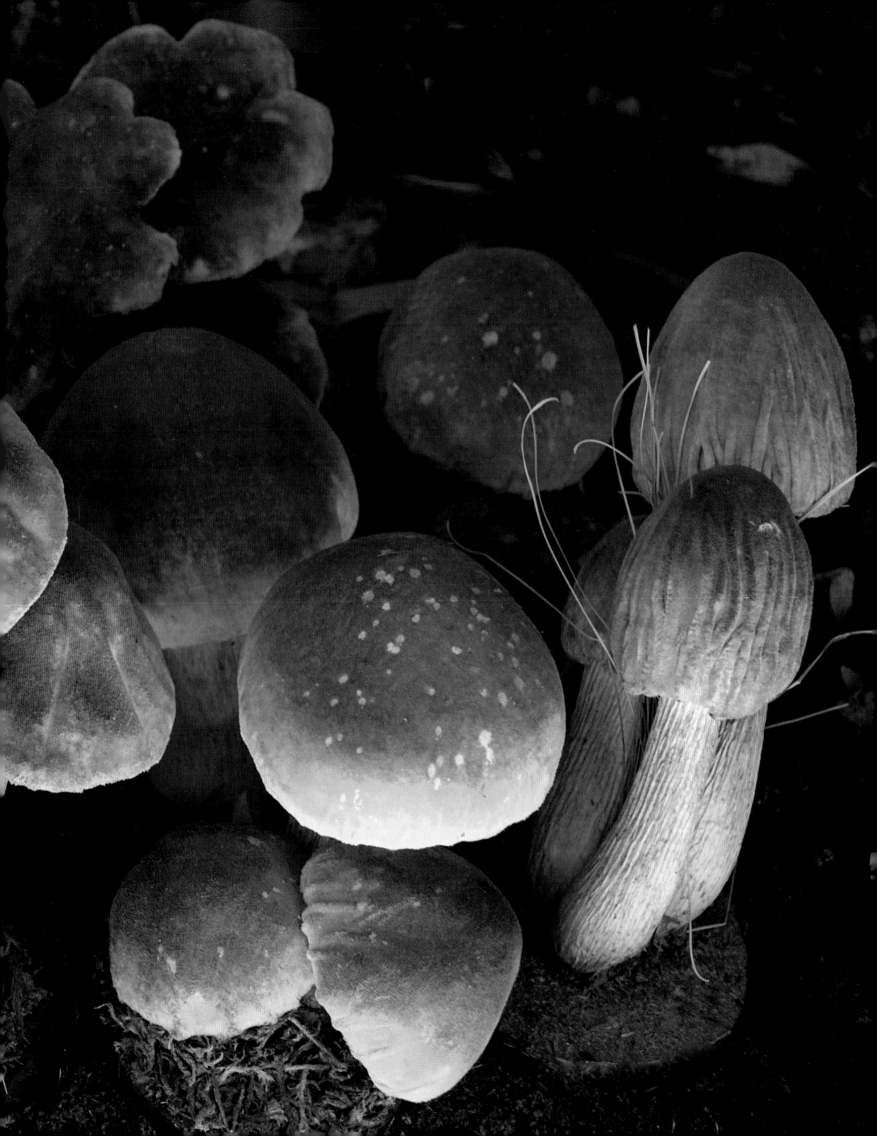

From *Atlas der ophthalmoscopie*, SEE P.150
OVERLEAF Models of pomegranates, SEE P.237

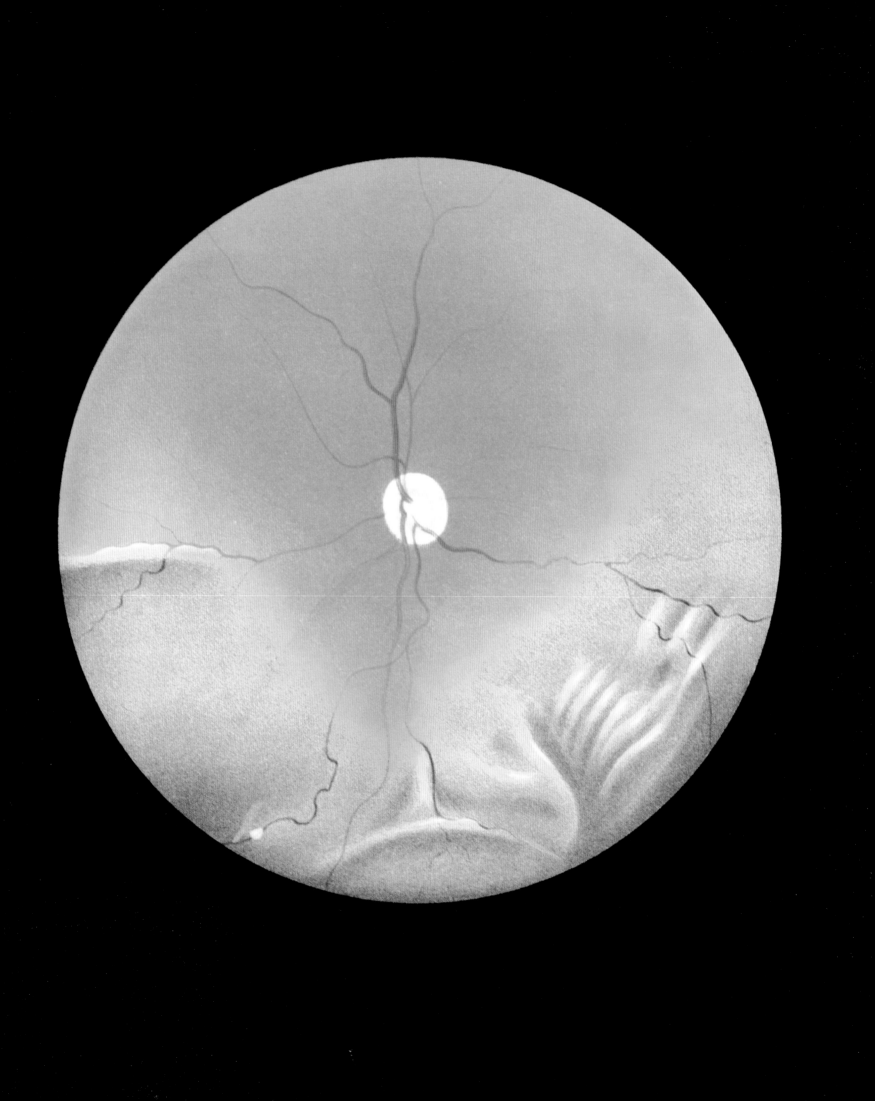

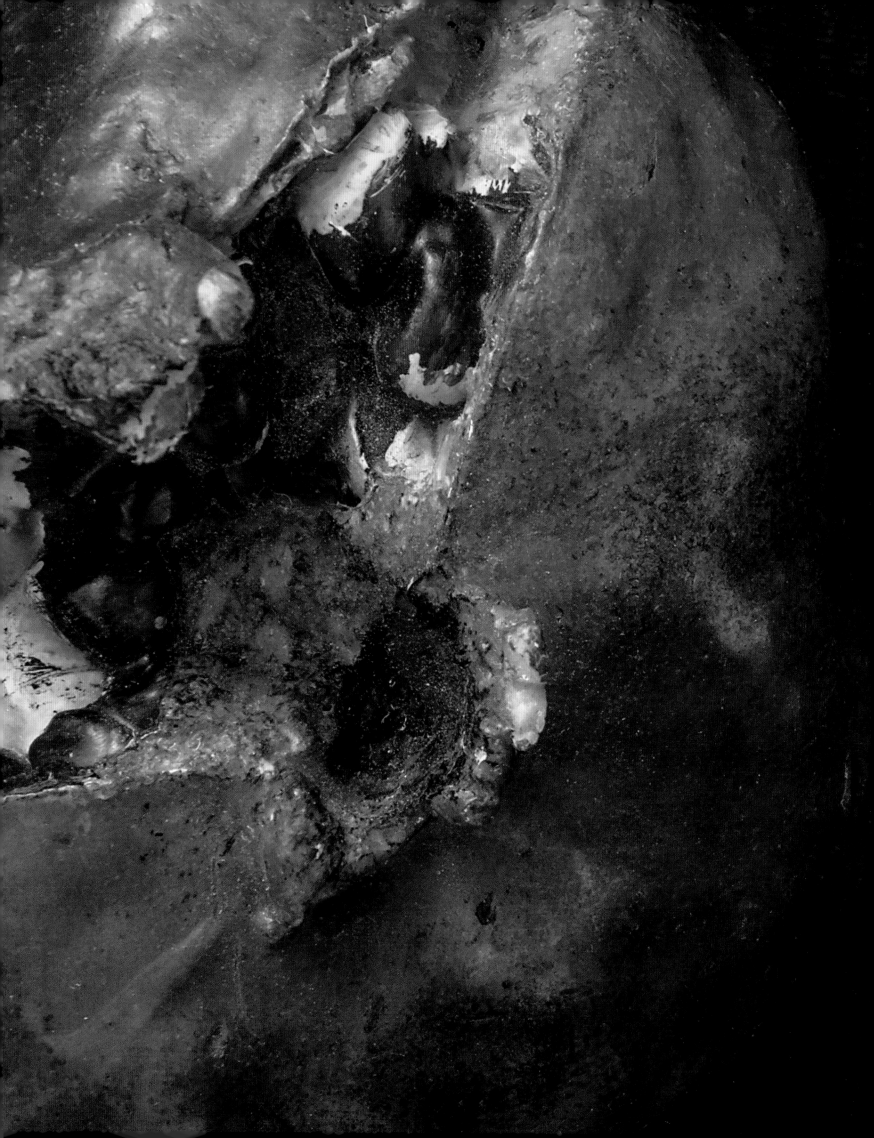

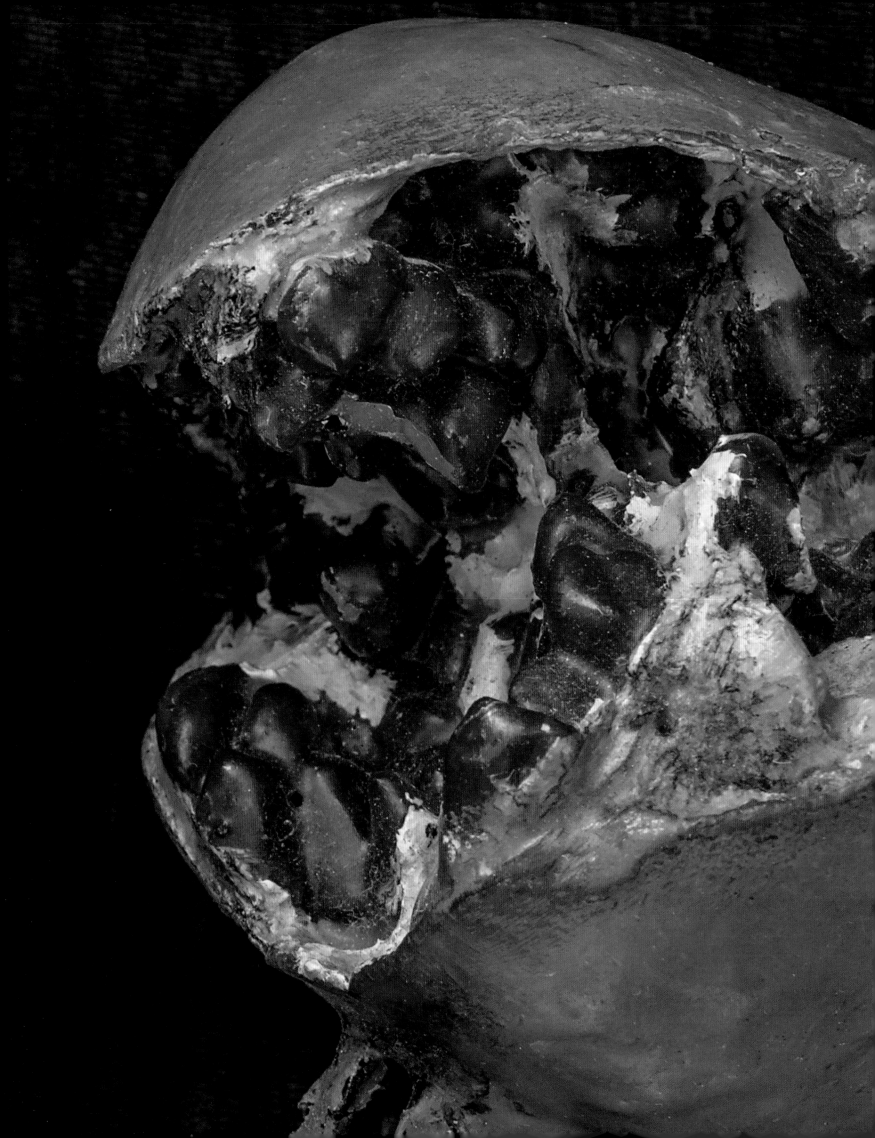

From *The wonders of the microscope,* SEE P.132

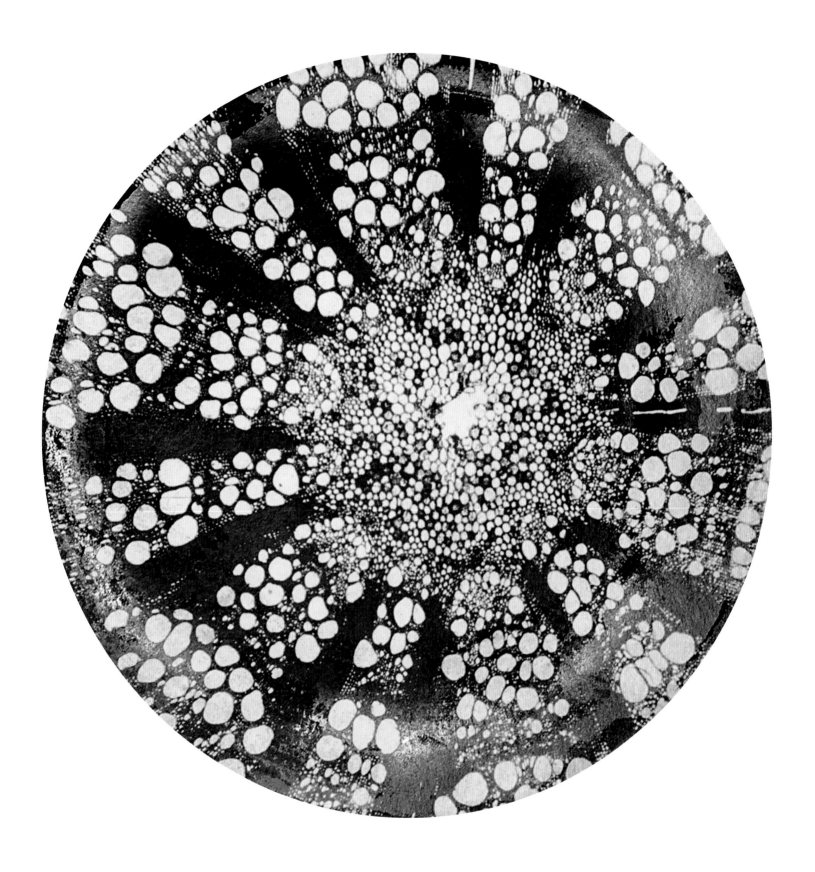

From *Recherches sur les poissons fossiles,* SEE P.235

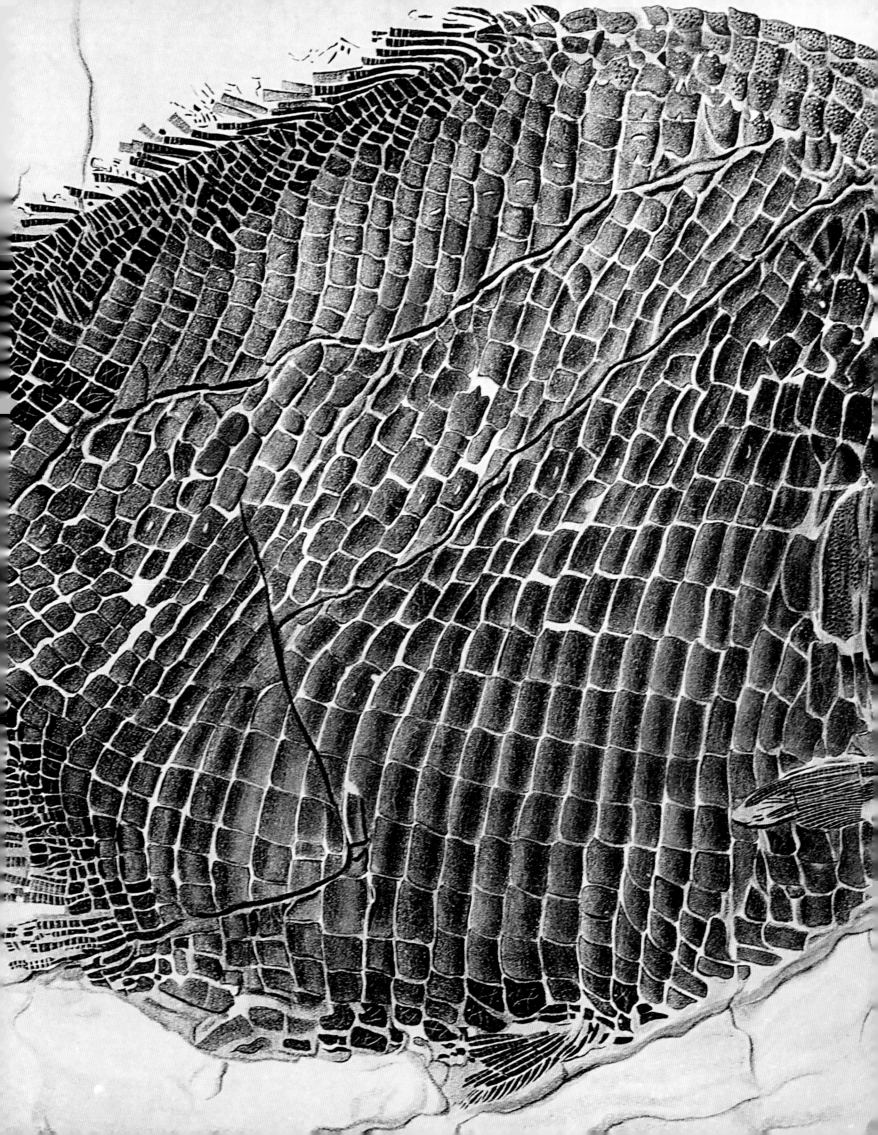

Model of jellyfish, SEE P.227

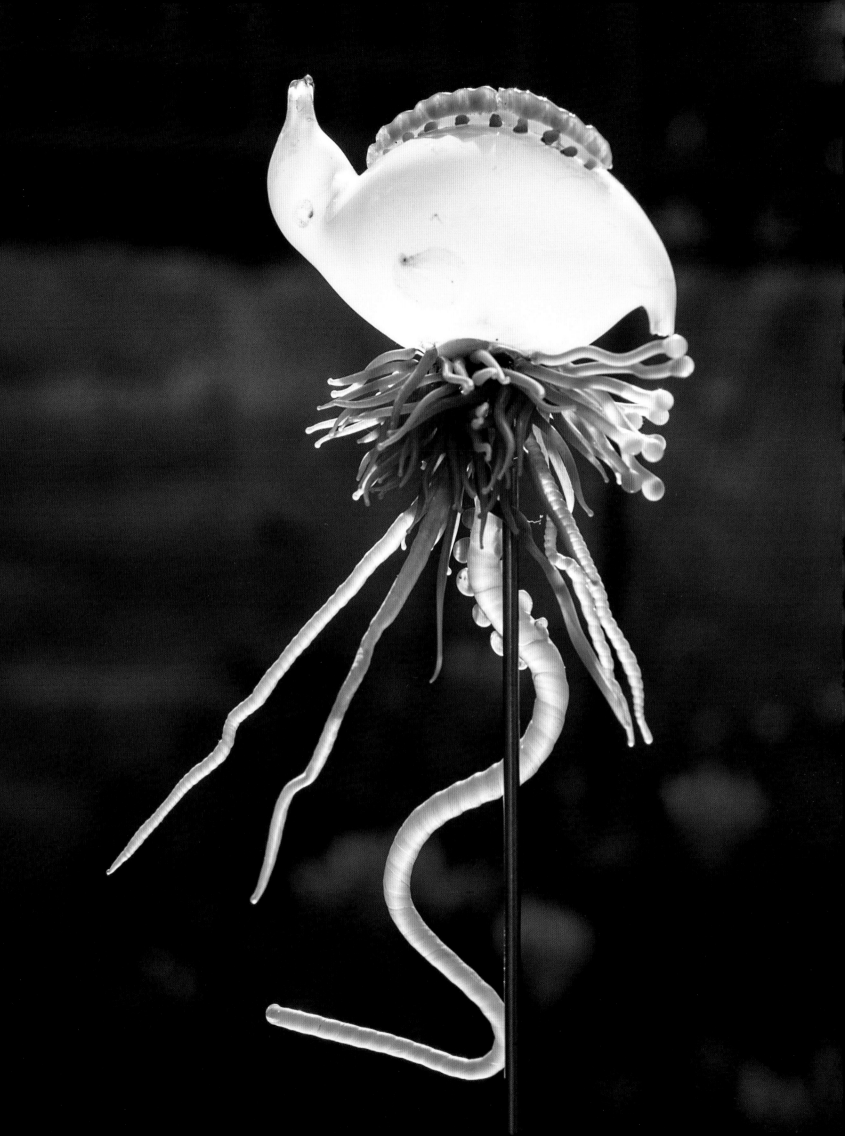

Object Lessons

Lynne Cooke in conversation with George Loudon

Lynne Cooke: Let's start at the beginning. You have made art collections in the past. What took you in this very different direction?

George Loudon: What took me in this direction was that I collected contemporary art for a long time and after 25 years I sort of ran out of puff. I had 700 works by 250 artists. Once I found a place to show it all – I actually had it all up – and that was somehow a final point.

In the meantime, I had begun thinking 'I don't know much about science' – so I started reading about it. I had seen artists who were playing around with science, playing around with certain aspects of it, or going back to certain nineteenth-century aspects of it. The combination of reading about it and seeing that some artists were doing things that related to eighteenth and nineteenth-century science, was reinforced by travelling to South America. I read about naturalists like Charles Darwin, Alexander von Humboldt, Alfred Russel Wallace and Henry Walter Bates, who all had been there. Then, because my daughter and son-in-law were at Harvard, I saw the glass models by Leopold and Rudolf Blaschka at the Natural History Museum there. These glass models were really an eye-opener in terms of what this kind of material could mean. So then I started looking everywhere, and going to an endless number of museum storages because, of course, the institutions don't show this material anymore. It's all hidden away. That's how I started: I bought what I could but there was not much around.

LC: As it's developed into its present form today, the collection has taken on fairly clear parameters. As with any collection, the exceptions help to show what the core is…

GL: It's primarily late eighteenth and nineteenth century, with the odd exception where I feel that going a little bit further in time allows for some of the characteristics that I'm looking for to be drawn out. The period that fascinates me is the time of the explosion of science.

But the collection is really, I think, about the pleasure or the 'wonderfulness' of these objects and illustrations, which originally were fairly commonplace. They were not made as collectors' pieces or museum pieces, or for cabinets of curiosity. They were made for very down-to-earth teaching purposes. Now that they've lost that function we can do with them whatever we like. I think the collection has something, because a huge amount of it was made by very clever craftsmen and women at a moment of great scientific expansion.

> **LC:** The moment that science explodes and there's a need for pedagogical models precedes industrialisation and mechanically produced goods. It's also the moment when both the self-taught amateur and the professional scientist can contribute significantly to their disciplinary field, whichever one of the natural sciences they are engaged in. Take [John James] Audubon, for example, who dissects and sketches his birds in situ – in their natural habitat – without recourse to highly specialised technical equipment.

GL: Yes, indeed.

> **LC:** In addition, in the period your collection is drawn from, science is accessible to what might be called an educated public. Consequently, such models would likely be of interest not only to students in the specific pedagogical situations for which they were primarily intended, but to a wider lay audience.

GL: That's one side of it. The other side is that there were scientists (or 'professors' if they are in Germany) who had developed evidence-based theories about certain things. The pressure on them in the teaching environment was to provide visuals. You sense a reluctance to do that because they say, 'It has to be exact, and I can't be exact. I know that this is the case, but if you force me to make a picture of it, then I will be open to inexactitude.' That's true even when microscopes are used. Because microscopic photographs did not yet exist, somebody had to make a drawing from a microscope and the drawing was then turned into a printing plate. The critics then say, 'Yes, but there are too many slips between those stages.' So there develops a pressure to use photographs in science, which was already a stage better. Actually, you had that discussion some 20–30 years ago, when the medical profession was reluctant to digitise medical photographs and X-rays because they said detail could get lost. Now that's been resolved, but 20 years ago there was resistance because the truth is quickly lost.

> **LC:** All of this debate revolves around questions of translation and representation. For example, the Blaschkas translated two-dimensional drawings of invertebrate sea-creatures into three-dimensional glass artefacts. A three-dimensional model allowed people who didn't have direct access to these very fragile and ephemeral creatures better ways of

studying them than was possible with a two-dimensional representation. The question of what is gained or lost with each medium is an interesting one. When the great compendium of American folk art, *The Index of American Design* (1935–1942) – an endeavour to record all the most important vernacular artefacts in the US state by state, genre by genre – was planned in the mid-1930s there was a debate about whether the objects should be recorded photographically or by means of watercolours. After consultation with archaeologists, the decision was ultimately taken in favour of watercolours because it was argued that they could provide certain crucial types of information that a photograph could not. Of course, the craftsmen and women who were to execute them had to be taught a certain methodology because the kind of 'truth' that was sought was of a particular order, involving a certain set of stylistic and formal protocols. Many of the works that resulted are not only important as documentation, but are also aesthetically very striking – very beautiful.

GL: In palaeontology today there's still a commitment to the practice of drawing because drawing can reveal nuances of fossils that a photograph cannot. Now, of course, with digital imaging, you can scan each layer and increasingly produce a much better sharpness in your picture. This whole issue involving reproduction and facsimiles is very much talked about now.

LC: It seems directly related to your field because, in looking across what you've collected, there are models which appear close to diagrams, there are watercolour representations that are reprinted as plates in books, there are three-dimensional one-to-one scale simulations (in velvet!) of mushrooms.… This panoply of materials – from lost wax casting, to direct nature-printing, to cyanotypes – is put in service to a vast range of different kinds of information. Within your collection, we find artisans inventing techniques or moving into areas where they have to think freshly about how to code particular forms of information.

GL: There is indeed quite a lot of experimentation. Nature-printing, for example, was such a constant mode of experimentation – and not least butterfly nature-printing, which was a fairly absurd way of doing it, but they tried! There's one example, which I don't own but which I refer to, of somebody who successfully nature-printed a book of butterflies that required the use of some 50,000 specimens. It seems absurd!

LC: That brings up the question of the representative example as opposed to the unique specimen – and the range of options along that spectrum. Presumably your artisans are looking to select representative types. Although you have stated that you're not interested in the *Wunderkammer* and so have tended to shy away from the unique (or freak) object, the weird or fantastical, the collection does include the Siamese...

GL: Siamese piglets and the two-faced kitten.

LC: They seem rather on the edge of things.

GL: Yes. They are on the edge of things – but they have an aesthetic. I think the way I went about collecting was very much a continuation of the way I was looking at contemporary art – I think what I've done is very visual. I mean to say, all my choices have been very visual and then the storyline (if there is a storyline) comes afterwards. And there's no systematic approach – it's one after the other – there's no attempt at grouping. I resist grouping in the book too. Each object has to present itself and it's largely a visual story.

LC: So that would be part of what distinguishes the collection as formed by an individual – one person's collection – from what a professional would do who was making a collection for a museum devoted to the history of science?

GL: Exactly. I would argue that they would have to include things that I would not find visually appealing.

LC: But there is some cross-over because there are craftsmen who are exceptionally skilled creators and who nonetheless made central contributions to a field, as the Blaschkas did...

GL: Maybe this is not totally germane to what you are asking, but I don't really go for scientific instruments.

LC: Rather than instruments, you focus on pedagogical models?

GL: Yes, I'm really dealing with material that is trying to convey content of some sort. It's visual, it's wide-ranging, but contained within certain parameters. There's not much for sale, though there's a lot in storage.

LC: One of the things that strikes me when looking at the collection is that it represents what could be described as a distinct field, one which, I think, has barely been identified as such. But before we get into that I want to ask you why there is not much more material for sale. What happened to it? Is it still in attics because people have not realised there's a market for it; or was it lost, or destroyed?

GL: I think that not much of it was ever in private hands and most of that was lost. Since the greater part was made for an institutional market, it's therefore stuck. It's either been thrown out of universities, scientific museums and schools or it got put away in a cupboard. Now they're reluctant to sell because they've been told it's not ethical. There's no de-accessioning – at least, in Europe, there's very little. Mostly, it's in storage.

What I find amazing is that if I go to any natural history or science museum, inevitably whoever is with me, usually the curator, is always surprised by the things that I like because he or she usually doesn't get it. There's a blind spot about this material. In most museums there's an implicit sense that everything in storage is second rate because, if it weren't second-rate, it would be on show. If, one day, museums do something with this sort of stuff I think they'll have to bring in outsiders. They will have to bring in the Rosemarie Trockels or the Cornelia Parkers of this world because those artists have got an eye, whereas the people who are responsible for those collections, with some exceptions, haven't.

LC: This brings me to the question of identifying a field for study. I would compare this situation in some respects to the moment when, immediately after the First World War, artists in the US started to 'discover' and collect nineteenth-century folk objects from the Northeast – artists like Alexander Calder and Marsden Hartley found in traditional folk objects, such as whirligigs, weathervanes and shop signs, stylistic affinities with their own work. Inspired by their 'discoveries', they in turn advocated for them, promoting folk art within a larger cultural movement then preoccupied with the question of what was an American art. Objects, which had not previously been collected – and were hidden away in barns and attics – started to be acquired first by artists and then by dealers. Within a decade, Abby Aldrich Rockefeller, who was a leading collector of contemporary art and one of the founders of the Museum of Modern Art in New York, had also amassed the foremost collection of this material. In short, within a very brief time, it segued from the periphery, so to speak, where it had no value and no visual interest, into the Museum of Modern Art as revered aesthetic objects.

In a like but slightly earlier case, European modernist artists – including Picasso and Matisse – started looking at African carvings and, as a consequence, those objects moved from the ethnographic into the contemporary art arena and even, for a time, into museums of modern art. In the course of that transformation, judgements about which pieces were formally remarkable, and hence of exceptional aesthetic quality, were made by leading critics like Roger Fry and others. Do you think that a parallel transformation might attend your material?

GL: In the examples you give, the objects were completely removed from their original context and then were viewed through new eyes and so the taste, or the aesthetics, of the viewer was imposed on those objects. That's exactly what I'm doing here. The only difference is that I don't attribute any permanence to this; I don't make any claim that this material is going to be as interesting in 50 years' time as I feel it is now. In 50 years' time, maybe somebody like me will be interested in completely different things. So, I think quilts and other things like that may be less interesting then.

LC: Responses to objects that lie outside the canonical art arena certainly do go through phases, yet certain judgements about quality that result in appropriation into the realm of fine art do endure: certain tribal objects from Africa are still today, almost a century after Picasso and his peers first engaged with them, considered aesthetic masterpieces equal to those of the greatest of cultures of any era.

GL: Yes.

LC: If you are indeed a pioneer in identifying a new field for aesthetic consideration, then this in part depends, I would argue, on the fact that your field has clearly defined parameters and criteria. As we discussed, it's defined by science at a particular historical juncture, science comprised of fields like natural history and so forth; by the objects' utilitarian roles; by the skill-sets of the trained craftsmen needed to make these works; and by an underlying aesthetic predicated on directness, formal inventiveness and economy. That nexus of factors dissolves or disappears after a certain moment. Science goes in different directions, the models are made differently and the period, as defined, ends.

Although you seem to be one of the few people who, to my knowledge, collect in this area, artists have responded to this material, albeit largely intuitively rather than in the terms that you identified. A parallel can again be drawn with the two groups of 'discoverers' responsible for the historic examples I gave: in those instances too, it was artists and a few enlightened collectors who led the way.

You yourself may, in due course, move on to something else, but the way in which these types of objects are of great interest now is revealing; it's indicative of something larger that touches our time. As you say, we can identify a slew of artists with different aesthetics, from Diana Thater to Rosemarie Trockel, Roxy Paine and Mark Dion, who share an interest in this material even as they respond very variously to it.

Having identified a field, collectors, critics and scholars typically go on to determine the canonical examples, the rarities and exceptions, and the more aesthetically charged items. In due course because objects have 'social lives' so to speak, the aesthetically charged examples may be drawn into the broad arena of art, an arena which encompasses vernacular and applied arts, folk and much else, and so make a transition from their former fields or disciplines of ethnography, anthropology and their subsets to the purview of art. This does not necessarily mean that these objects cease to function in their primary arenas, the scientific fields that formerly held sway over them, but rather that as artefacts they can be considered through different disciplinary lenses. Bringing them into the domain of 'art' doesn't entail embracing them as works of 'fine' art; there are many other categories of objects we call 'art' beyond those of painting and sculpture as practiced in the Western world since the Renaissance.

GL: I suppose my objection to calling this material 'art' or even allowing it to be presented as such is that it was all made to be looked at from very close by, not behind glass. A lot of it is to be touched, or taken apart, and looked at up close, sometimes in a book. It's very democratic – it's populist stuff – and as soon as it's put into a museum, it's contextualised in ways that attribute an effect of grandeur to it. If there is a guard standing by so that you don't get too close, that would kill it – well, maybe not entirely but largely. So though I think natural history and science museums should have tons of this stuff, they will have real trouble showing it because they're going to apply the same conservation standards as they apply to a dodo bone. They're going to put it in expensive, climate-controlled cases. That's my problem with the prospect of this becoming museum art. I've thought a lot about how you could show this material and, for me, the Wellcome Trust in London is ahead of everybody else in experimenting with how to display material of this kind.

LC: There are several issues here. The first is whether having lost their original function, these objects can only retain a sense of their roots if they are kept somehow under the purview of those sciences for which they were originally destined. We seem to agree that that seems unnecessarily limited: this material can enter other disciplinary discourses without misrepresentation and without forfeiting all reference to its past lives. Then there's what art historian Svetlana Alpers calls 'the museum effect', which relates to the question of display. There's no way of looking at any of these objects outside a frame – whether it be an institutional frame or a conceptual frame: when they enter the public arena we address them in terms of various predetermined intellectual schemata. Alpers argues that when any object, natural or man-made, is put on public view in a museum of whatever kind, we are indicating that the object has visual distinction and also cultural significance. If I understand you correctly, you want to play down claims to cultural significance and foreground aesthetic qualities. When considering these objects in terms of their visual distinction, we start to look at them formally, technically, materially...

GL: With a bit of history attached.

LC: A history of the object's making, or a history of its purpose?

GL: Both, I think. I think you have to explain a little bit about why it was made – but not too much – and, if you know, who made it. Take the Blaschkas. At Harvard the glass botanical specimens are still presented in the cases that were made for them in a fairly traditional late nineteenth-century, early twentieth-century presentational setting. In Dublin at the Natural History Museum, they are also displayed in that way because they're still used to demonstrate taxonomical knowledge: there you may have an invertebrate next to something with a shell – the actual shell – and the invertebrate is represented by a Blaschka glass model. The labels that give the name of the specimens in Latin now also say: 'Blaschka'. So today, the makers, the craftsmen, of the particular works are identified whereas 20 years or so ago they were not. The objects were treated simply as specimens.

In 2002 the Design Museum in London had a show of Blaschka works together with sculpture by Dorothy Cross, a well-known contemporary Irish artist. I felt that it didn't mix all that well. The Blaschkas' models were presented beautifully in very simple glass cases and illuminated like pieces of jewellery. It was quite spectacular. I thought that they worked quite well.

LC: When you juxtapose Blaschka objects with works by Dorothy Cross, you invite a certain kind of dialogue that's very different from the discourse set up in the Natural History Museum.

GL: I would rather see Cornelia Parker curate a show that included works of this kind than an exhibition that seems to say: 'Hey! I've included a contemporary artist just to make you think that this might be contemporary art.' In the recent exhibition at the Mori Museum in Tokyo of medical material from the Wellcome Trust in London, a couple of pieces by Damien Hirst were included, as if to say: 'You've got to look at this material the way you look at a Hirst.' I would like to see a little more sophistication.

LC: In some of these cases artists seem to have been commissioned to curate shows with such material because their imprimatur will make us take it seriously. I don't think we need to go that route any longer; we can contextualise individual objects of this kind more richly, within wider thematic and more exigent relations. However, we may need to distinguish between incorporating individual artefacts and dealing with extensive collections such as that of the Wellcome Trust, which was originally a private collection, amassed on a vast scale, primarily for pedagogical reasons. It was not selected with the kind of attention to aesthetic concerns that you've brought to your collecting. When we look at individual items there, many seem to function as representative examples of, say, pipes, or of medical instruments for birthing, and so forth. Nonetheless there are high points, when the object is also aesthetically compelling, and others when the object is historically significant. The Wellcome Trust has what could be described as a crossover audience, that is, visitors with wide-ranging areas of interest and specialists, historians and the like.

GL: Yes, exactly.

LC: Would the Wellcome Trust be the ideal site for presenting your collection?

GL: Yes, in theory; no, in fact – because they are committed to thematic exhibitions. I'd like to see material like mine shown in a contemporary art venue like the Hayward Gallery in London or the Drawing Center in New York, which several years back did an exhibition called *Ocean Flowers* (2004) which, in terms of subject, was, you could argue, somewhat off their normal track.

LC: Yes. Given that the show included cyanotypes and nature prints, the material was 'proto-photographic' yet still within the realm of drawing, broadly defined, and so was a wonderfully inventive response to their mission, which, unusually, is medium-based. While the Drawing Center mainly shows the work of contemporary artists, it does, occasionally, range widely. Shows like *Ocean Flowers*, which position contemporary drawing relative to forms of drawing historically and sometimes also culturally far removed, energise and complicate their programme in challenging ways, prompting us to rethink our familiar ideas about drawing as an art form, its boundaries and its potential. Presenting your material in such a context might encourage viewers not only to engage with it in primarily aesthetic terms but to think about the intersection of art and science today, and the function of visual models in each arena.

Your proposal also raises another question. Do we have suitable institutions for displaying your material? Can we imagine institutions that would specialise in it – in the same way that, following the upsurge of interest in the US in the early twentieth century that folk art generated, initially by artists and pioneering collectors, museums specifically dedicated to that material were then founded. Today institutions that engage with historic and contemporary folk art can be found across the country. Could we argue that yours is a burgeoning field that is only now beginning to gain recognition but that, in due course, museums might be founded to address it?

GL: I find it difficult to imagine that, but I haven't really thought much about it. I suppose one reason why I haven't thought about this much is because I have a slight aversion to institutionalising these things. Today you have people making collections of dust and sand and as long as you put it all in identical little bottles, it can become a collection. There is an institution here in London called The Museum of Everything, which takes the view that you can make a collection of anything as long as it is properly curated, and visually attractive. I think that approach has become a sort of movement, one in which some institutionalised museums participate as well.

LC: One problem with your material may be its scarcity: is there enough in circulation to make a market, develop a constituency and hence a social presence that would encourage the amassing of collections? If that's not an option, then the alternative would seem to be to prompt institutions that have this material to think differently about it.

GL: Yes. If you show the botanical teaching diagrams, which belong to the Whipple Museum in Cambridge, by John Steven Henslow – Darwin's botany teacher – at Kew Gardens, on the whole people will walk right by them. They won't look for more than a second. If you hang them in the Serpentine Gallery in London people will say, 'Hey, what's this?' I feel this material needs to be shown out of context.

> **LC:** Having legitimated its aesthetic value by contextualising it within the sphere of visual arts institutions, could you then make an argument to natural history museums or to science museums, that they need to consider it differently, not merely to establish a section of the institution devoted to the histories of science but to rethink its cultural image in terms of the 'modern eye'? At present those institutions either don't think in those terms or they don't have the resources to do so.
>
> For the moment, however, you have decided to publish a book, and tellingly, to publish with a press that's known for fine art publications.

GL: I am pleased that Ridinghouse is publishing this. I think that amongst people interested in contemporary art of a certain nature, there are quite a few that would be interested in this material. Among the reasons for this – though I'm not quite sure of the order of importance – is that when people who see a lot of contemporary art see this material, they can see some links. They can see links to artists who are inspired by the same subjects, or they can see how, if taken out of context, it can be used creatively to new ends. Also, I sense amongst contemporary art aficionados a certain nostalgia, which may come from a relentless bombardment of...

> **LC:** Of the spectacular?

GL: Yes, a reaction against Jeff Koons, if you will. So that's the third reason. Then the fourth reason is that, I feel there's a vacuum at the bottom of the contemporary art world. Contemporary art museums are becoming very strong in their pronouncements about what is good and what is not good contemporary art.

> **LC:** Do you mean that today they show little that is explicitly alternative and experimental?

GL: Yes. And the galleries are following suit. They're getting bigger and bigger and the works they show are more and more museum-sized. Sometimes they'll even borrow art works from museums which serves to give them the same sort of implicit authority as the museum. They can say, 'Well, you can't have that one, it belongs to *x* or *y* museum, but you can have this one.' I think quite a lot of people are fed up with this situation and are looking underneath. They're looking for quirky, curious and strange things that they can look at intimately. The objects I collect fit really well that need.

LC: I would add to your list of reasons a growing interest in skilled craftsmanship. I think the fact that these models are not only made by hand but sometimes with an extraordinary technical inventiveness, as was the case with the Blaschkas (we still don't know exactly how they did them), or with great refinement, as in the making of the box of peaches with deformities, is also telling (p.220). The impact of the hand (a sense of the hand finding inventive ways to represent something hitherto unknown with familiar materials), and the investigative attitude (because these hadn't been made in that way before) is very evident and appealing.

The requirement, that they demand, of close scrutiny returns me to Svetlana Alpers, who begins her essay 'The Museum as a Way of Seeing' (1991) by talking about a childhood memory of looking at a very large crab in the Natural History Museum at Harvard: an act of paying heightened attention to what she saw, its individual components, and to how all the parts came together. Something of that absorbing experience, of being in immediate touch with the natural world, occurs with your pedagogical models. Their subjects are mostly related to our bodies or to our physical environment, and these two subjects are of course of immense and urgent importance.

GL: Yes, that makes an enormous amount of sense.

LC: In addition, and maybe this is related to what I was saying earlier, the models stem from and depend on acts of direct perception and empirical discovery, even when new tools like the microscope are employed... By contrast, the visual models made for many branches of science today are deployed to image data. In some fields, like astronomy, scientists have recourse to models ranging from two-dimensional diagrams to four-dimensional computer generated imagery, in order to get to grips with vast quantities of complicated information. What results doesn't correspond to something that could actually be seen with human eyes. Moreover, these visual models are often highly technical and so not easily read by a non-specialist; or they have to be translated by means of a lot of background text before we, the general public, can comprehend. As a result, I think they are of a different order from yours, and don't generate the kind of fascination we feel with those artefacts, like the velvet mushrooms, that have a one-to-one correspondence with actual objects in the world that we see and know.

GL: That makes sense.

LC: Given that your collection was made by one individual it inevitably betrays your taste, including its quirks. To take a couple of examples: there is a map marked with the reservations on which Native Americans were confined in the nineteenth century (p.99). Another example of an 'outlier' would be the arrowheads, which I believe you said were found in South America (p.90). Then there's a book with drawings of different states of the human soul (p.177). How do these mavericks, if you agree that they warrant this term, function within the framework of the collection?

GL: A quite simple little piece of paper, the map covers a huge piece of history with amazing aspects to it. When the Native American tribes who wouldn't sign up to integration were sent packing in the 1830s, some of them had their own slaves: because there was slavery on the Eastern Seaboard, the Native Americans' slaves were sent along with them! This map has all these curious human stories attached to it. The tragedy of these Native Americans was that they were sent marching with nothing – just what they could carry on their backs. It's such a strong story – an abomination really – it brings tears to the eyes.

The arrowheads are from southern Argentina, Patagonia – from the Tehuelche, who were nomadic tribes. Their history is not unlike those of indigenous peoples in the US. They were increasingly contained and subjected to all the illnesses that the Europeans brought; and there were Indian wars very similar to those in the US. These arrowheads come from an area which wasn't really colonised until the late nineteenth or early twentieth century, and then very sparsely. The Tehuelche by that time weren't using arrows, of course. These examples are difficult to date because there are no stylistic clues. They're not very abundant and are found where they were made – in places where the stone was appropriate. In addition, there's a link to Darwin who travelled extensively in those parts. He met the Indians and dealt with them. I like that part of the world a lot because it's so empty and so strange – so I quote a bit of Darwin in the text when I'm describing the arrowheads.

The soul book I found in a bookshop in Los Angeles is another extreme within the parameters of the collection. It was just so strange: it had illustrations of differently shaped and differently coloured souls by a writer who claimed to be able to see people's souls. There is a famous book from the 1950s – *The Third Eye* (1956) – about a Tibetan who had a hole in his skull, which allowed him to see people's auras. This is rather like that: it describes how the author can read people's souls by the shapes that hover over them. I thought it was quite enchanting with its very pretty little prints.

These examples do lie on the extremes – but there we are.

LC: Have you at any time bought to fill 'holes' in your collection?

GL: No, I don't fill holes. There are things that I missed in the sense that I wasn't quick enough, or I was outbid, or whatever – but 'holes' implies that you know the field and I don't really know the field.

LC: Do you have a wish list?

GL: No. There are things that I've seen which I would have liked to have.

LC: But that's in retrospect.

GL: And if I see something like it again, I might... but I'm not even sure I'm going to add very much to the collection now. I'm more interested in what you can do with this sort of material than in expanding.

LC: Is that because of the practicalities of dealing with the collection's future?

GL: It was the same when I was collecting contemporary art. The practicalities at a certain stage become overwhelming. I remember buying works I couldn't get through the front door. Then I said, 'Wait a minute. Now this is going too far.' And then I had to stop. There are wonderful things out there – occasionally I see an amazing wax *écorché* (a sculpted figure showing the body without skin) – but they're very, very expensive now. What I have is not expensive.

LC: Well, that may be because in large part you have been ahead of the game.

GL: Yes, maybe so.

Looking Again

Robert McCracken Peck

The remarkable assemblage of natural history specimens, scientific illustrations, books, museum models and didactic teaching tools that are the focus of this book confirm Mark Twain's adage that truth can be stranger than fiction. Truth can also be more beautiful. Although all of the artefacts shown here have the common purpose of trying to represent or interpret the natural world, they reveal that nature's 'truth' can have many aspects, and can be at times seductive, inspiring, disturbing, disorienting, and even surreal. While some of these models and illustrations are straightforward representations of nature, others are so alien to our eyes that we might assume them to be the expressions of artistic visionaries from a later time, rather than Victorian interpretations of scientific fact.

The nineteenth and early twentieth-century scientists and educators who commissioned these works – and the artists who constructed them, one can assume – had the same objectives as their modern-day counterparts: namely, to encourage audiences to look carefully at nature from new and original perspectives. Sometimes they achieved that novelty by creating a change in scale, or by taking a familiar detail and presenting it in an unfamiliar context. At other times they did so simply by looking at something closely and showing it accurately for the first time. We now have the opportunity to look at what they created again, not as didactic tools through which specific ideas were once conveyed, but as intriguing artefacts and images worthy of attention in their own right.

Each object or illustration reproduced in this book, no matter how beautiful or utilitarian, was crafted to help students understand or interpret some aspect of life on earth. There are even a few that represent worlds beyond. Those created from a pre-Darwinian perspective were often conceived as a way of appreciating God through his creations. Others were more mechanistic and practical in intent and address scientific questions such as: How are plants fertilised?

How is the human body built? The materials used to create them – wood, metal, glass, papier-mache and wax – have more to do with practicality than artistic expression, and yet the disparate techniques used in their fabrication help shape how they affect our sensibilities today. These now often-ambiguous objects were intended to help people better understand the complex structures and organising principles of the natural world. Today, in the context of a society in which shape, form and surface appearance may trump intent, meaning and function, they are open to new interpretation.

As marketing and advertising firms demonstrate with unremitting regularity, the way in which something is presented can make an enormous difference in how it is perceived. This is as applicable in the field of natural history as it is in any other. Consider, for example, the controversial installations by contemporary British artist Damien Hirst in which whole animals are suspended in vats of formaldehyde or alcohol. In works such as *Mother and Child (Divided)* (1993) for instance, the artist presents the two preserved halves of a cow and calf in four glass-walled tanks. These displays of bulky biomass command critical attention and large sums of money from both private and institutional art collectors, while identical specimens, created as teaching tools by anonymous technicians and displayed in less appealing containers are ascribed a much lower place in the hierarchy of desirable artefacts. As a result, they have a substantially lower monetary value (if they can be acquired at all) and may even require special licences to own. While the Hirst pieces are eagerly sought out and admired as original creations, the teaching cadavers are shunned by all but veterinarians and biology students obliged to study them in order to earn their professional degrees.

For as long as we have studied nature, we have tried to arrest its decay. By immersing specimens in alcohol, dusting them with pepper or tobacco, rubbing them with arsenic, or encasing them with resin, we may slow, but never entirely eliminate the effects of time and insect predation on organic samples from the natural world.[1] For this reason, seventeenth-century keepers of cabinets of curiosities sought out physically stable objects such as skulls, horns, antlers, bones, fossils, minerals and shells for display. They also collected the skins of armadillos,

1 For a detailed account of the various methods used to preserve specimens, see Robert McCracken Peck, 'Preserving Nature for Study and Display' and 'Alcohol and Arsenic, Pepper and Pitch: Brief Histories of Preservation Techniques', in Sue Anne Prince (ed), *Stuffing Birds, Pressing Plants, Shaping Knowledge*, American Philosophical Society, Philadelphia, PA, 2003, pp.11–25 and 27–53.

alligators and puffer fish, all of which could be easily dried and preserved. The practice of taxidermy came later, but without much lasting success until the first half of the nineteenth century.[2] Specimens preserved in rum, brandy or other spirits lacked aesthetic appeal, and though scientifically useful, have never attracted the same degree of affection by amateur collectors as more stable and more colourful examples of natural diversity.

Due to their fragility and fugitive colours, pressed plants were at a similar disadvantage. Whether kept for their sentimental value, or collected encyclopedically by serious botanists, they were rarely featured in the natural history cabinets that were once so prized by well-heeled amateurs.[3] Not until the nineteenth century, when ferns, seaweeds and flowering plants became the subjects of popular study and were preserved in presentation albums or reproduced as nature-printed 'photograms', did these subjects gain widespread acceptance in collecting circles.[4] For those interested in teaching botany, larger-than-life models became essential tools for explaining concepts of anatomy and identifying diagnostic features in different species. Today, these objects, along with scaled-up models of insects and other organisms, assume new roles as semi-abstract sculptures. Was Lewis Carroll thinking of such oversized teaching props in the classrooms at Oxford when he changed Alice's scale with a bottled potion and a piece of cake during her adventures in Wonderland?

Whether they were housed in the *Wunderkammer* of seventeenth-century noblemen or in the Victorian albums and cabinets of nineteenth-century collectors, natural history specimens assembled and preserved for private contemplation or public display were valued as examples of nature itself. George Loudon, the twenty-first-century collector and author of this book, enjoys the beauty of nature in its wild state, but what interests him even more are the things scientists and artisans have created to interpret the natural world. His collection celebrates the books and three-dimensional models made to record, study and explain anatomy, natural processes and the mysteries of life on earth. Through the lens of these creations, and with the help of photographer Rosamond Purcell, Loudon offers in this book a unique perspective on the intersection of science and art. By exploiting the blurry lines between fact and fiction, past and

2 The earliest surviving taxidermied specimen is an African grey parrot that was buried in Westminster Abbey, London with its owner in 1702. See Pat Morris, 'The Antiquity of the Duchess of Richmond's Parrot', *Museums Journal*, no.81, 1981, pp.153–54. For a discussion of early taxidermy techniques, see Christopher Frost, *A History of British Taxidermy*, privately published, Lavenham, 1987 and PA Morris, *A History of Taxidermy: Art, Science and Bad Taste*, MPM Publishing, Ascot, 2010.

3 The oldest surviving institutional herbarium dates from 1545 at the University of Padua in Italy, although a collection of dried plants may have been associated with the botanical garden at the University of Pisa even earlier. The Italian naturalist Ulisse Aldrovandi (1522–1605) had more than 14,500 specimens and 2,000 illustrations of plants in his own collection. For more on this see Paula Findlen, *Possessing Nature: Museums, Collecting, and Scientific Culture in Early Modern Italy*, University of California Press, Berkeley, CA, 1994.

4 Photograms are photographic images made without a camera, produced by placing objects on a photosensitive surface in the dark and exposing both object and surface to light.

present, real and artificial, he invites us to take a new look at objects designed to capture and describe the complexity of nature. These artefacts, some rescued from the storage bins of history, may no longer fill the roles for which they were originally intended, but they still have meaning and value as beautiful and interesting objects.

Paleolithic paintings and carvings of animals found in almost every part of the world may well have been humanity's first attempts to document and control nature. Whatever their purpose, they also represent the intervention of an artist's hand in experiencing, recording and perhaps attempting to explain the mysteries of life on earth. Medieval herbals and bestiaries represent another important step in recording nature's diversity. Although these frequently include religious or mythological ideas about plants and animals, their purpose was to construct an intellectual framework for the life-forms on which humans relied for good health and physical sustenance. Those efforts ultimately evolved into more complex and objective ways of recording, organising and communicating knowledge of nature. With some exceptions, the many publications about the natural world that emerged during the seventeenth and eighteenth centuries offered progressively more sophisticated systems for classifying and understanding nature. As printing and bookmaking became more common, these ideas and the images that complemented them could be shared with increasingly larger audiences.

In the late eighteenth and early nineteenth centuries, in order to meet the demands and affluence of a growing educated class, teaching hospitals, universities and other academic institutions began to teach students in groups, replacing the more individualised apprenticeships and educational experiences of earlier times. In turn, this new approach to education created a need for instructional models large enough to be viewed by more than one student at a time and durable enough to survive months, years and even decades of hard use in the classroom.

Medical models were among the first of this kind of teaching tool, obviating the need for human dissection and all of its associated ethical, legal and physical challenges.[5] The first models to reach North America were three life-size sculptures of a pregnant woman in various

5 As the use of human cadavers for teaching was illegal throughout most of the eighteenth and nineteenth centuries, sculptural substitutes were regularly employed for educational purposes.

6 For more on these models, see Peck, 'Illustrating Nature: Institutional Support of Art and Science in Philadelphia, 1770–1830' in Amy RW Meyers (ed), *Knowing Nature: Art and Science in Philadelphia, 1740–1840*, Yale University Press, New Haven, CT, 2011, pp.221–22.

stages of dissection. These three-dimensional models were cast in plaster from life (death), given a lifelike veneer of colour, and sent to Philadelphia's Pennsylvania Hospital from London in the 1760s by the enlightened Quaker physician, philanthropist and Royal Society member, Dr John Fothergill.[6] In Europe, similar teaching models of human bodies were created in wax by the descendants of professional model makers, some of whom had been crafting religious figures for crèches and shrines for generations. Some of the few anatomical sculptures of this kind that survive can be seen in museums in Rome, Florence, Bologna and elsewhere.[7] Over time, a wide variety of other materials were employed to create similar models, all in an effort to explain anatomical structure without the limitation of natural decay.

In the field of botany, long the domain of the physician and apothecary, the Swedish botanist, physician and zoologist Carl Linnaeus's revolutionary method of classifying genera and species by their reproductive organs shifted the emphasis from a largely aesthetic, nutritional or medicinal pursuit to a more purely scientific one. Now, for the first time, details of a flower's component parts – stamens, pistils and ovaries – became essential for the accurate identification of a species. Microscopic illustrations and oversized models made of papier-mache or wax helped botanists and botanical instructors to identify plant families and to explain reproductive processes in ways that the actual specimens of fragile fruits and flowers would never allow. Other fugitive organisms, such as mushrooms, were also represented by models, crafted and painted with such fidelity as to replicate their living prototypes at the peak of their short life cycles in every way, save permanence. Whereas plants tend to decay within hours of picking, these accurate models could be observed and appreciated indefinitely.

Collectors with an interest in nature but without the means to acquire or the space to store or display actual specimens have long turned to illustrated books, prints and drawings as an appealing alternative. Widely admired as beautiful objects themselves, illustrated books on natural history indefinitely extend the shelf life of the organisms depicted and described.[8] These paper collections also have the advantage of being able to compress time and distance by showing different stages of development in a single organism, or by comparing related but

7 Among the best-known sculptors of these anatomical figures are Lodovico Cigoli, Gaetano Zumbo, Anna Manzolini, Felice Fontana of Tuscany and Clemente Susini. A collection of some 1,400 wax anatomical models, created between 1771 and 1850, is housed in the 'La Specola' Museo di Storia Naturale (The Observatory' Museum of Zoology and Natural History) at the University of Florence. Other examples can be found at Museo Storico Nazionale dell'Arte Sanitaria (The National Museum of the History of Medicine) at the Santo Spirito Hospital in Rome; the

Nationaal Natuurhistorisch Museum (National Natural History Museum) in Leiden; the Museo delle Cere Anatomiche 'Luigi Cattaneo' (the Luigi Cattaneo Anatomical Wax Model Museum) at the Institute of Human Anatomy at the University of Bologna, in Bologna; the Museo Anatómico (Musuem of Anatomy) at the University of Valladolid, in Valladolid; the Medical School Museum at the University of Malta: and the Mütter Museum of The College of Physicians of Philadelphia.

8 While attempts were made to gather accurate visual representations of nature during the Middle Ages, the invention of movable type and the resulting explosion of printed books in the sixteenth century made that the first great age of visual encyclopedias. For more on this early period in natural history illustration see David Freedberg, *The Eye of the Lynx: Galileo, His Friends, and the Beginnings of Modern Natural History*, University of Chicago Press, Chicago, IL, 2002.

geographically disparate species on a single page. Unlike the models and teaching props that are only now being appreciated as works of art, the illustrations in these books have always been recognised as such. The eye-catching beauty of these plates has often led to their removal from the books for which they were originally conceived and to which they added essential content, in order to be sold and displayed as works of art in their own right.

In a sense, time has provided a parallel stimulus for new ways of looking at three-dimensional models. Ceramic models of human heads, which once helped phrenologists plot personality traits and aptitudes, now appeal to the very zones of imagination they were created to identify. Although the ideas they represent have been discredited, they continue to intrigue the twenty-first century viewer as artefacts. Similarly, plaster moulds of horses' mouths and feet now create as much visual and intellectual interest for laymen as they did for the veterinarians who once read them in a very different way.

Illustrations and models, while intended to serve as objective records of 'truth', are, in fact, artificial productions of a highly subjective process, dependent upon the knowledge, skill and idiosyncratic perspective of the individual artists who make them. Those models created to explain invisible biological processes such as photosynthesis or cell division, or prehistoric life – as seen in the images of 'deep time' by natural history artist Benjamin Waterhouse Hawkins and others – are, by definition, theoretical interpretations based on varying degrees of physical evidence.[9] While less speculative, models and illustrations that record actual specimens or geographical phenomena and events, such as the detailed illustrations that were used to docu-ment expeditions of discovery, still reflect considerable selectivity and individual interpretation. Wood cuts, engravings, etchings, lithographs and other two-dimensional printing processes each have their advantages and disadvantages when committing the 'truth' of nature to paper.

For as long as natural history books have been made, scholars have vied with one another to secure the services of artists with the skills to accurately reproduce nature. The illustrators' experience in selecting the scale, perspective and composition of their subjects, and their

9 For more on Hawkins's two and three dimensional interpretations of prehistoric life see Valerie Bramwell and Robert McCracken Peck, *All in the Bones: A Biography of Benjamin Waterhouse Hawkins*, The Academy of Natural Sciences of Philadelphia, Philadelphia, PA, 2008. For more on the depiction of deep time see Martin JS Rudwick, *Scenes from Deep Time: Early Pictorial Representations of the Prehistoric World*, University of Chicago Press, Chicago, IL, 1992.

ability to manipulate various printing processes and reproductive techniques, allows them to create images that are both instructive and visually engaging. Even when the subjects are not inherently appealing, the giant salamander in Philipp Franz von Siebold's *Fauna Japonica* (1833–50; pp.28–29) for example, the resulting images can be extraordinarily powerful.

Most of the objects in the Loudon collection were created for pedagogical purposes, but this does not in any way negate their aesthetic appeal. Consider for example the wax fruits and velvet mushroom models, or the stunning glass sculptures of marine invertebrates made by the Bohemian glass artists Leopold and Rudolf Blaschka.[10] All of these objects can be enjoyed as works of art, whether or not the viewer is aware of their intended use. Similarly, the exquisite illustrations of fish, shells and flowers contained in a group of nineteenth-century Japanese books, or the renderings of prehistoric fish fossils that forced skeptics to face the possibility of evolution, though educational in content, are all objects of enormous beauty.

In addition to classical books by some of the nineteenth century's most renowned scientists (including Louis Agassiz, William Buckland and von Siebold) the Loudon collection holds publications by many long-forgotten scholars whose books contain images ranging from telescopic studies of the surface of the moon to the wing scales of butterflies, and from the unseen fundus, or interior surface, of the human eye, to the eye-popping colours of the Northern Lights (pp.138, 223, 150, 192). One unusual book in the collection even attempts to chart the shapes and colours of the human soul (p.177). While the ideas and illustrations in the *Soul Shapes* (1890) volume may stray into the realm of the occult, the majority of the books in Loudon's collection were conceived as tools for recording and interpreting more tangible aspects of nature. Their goal was to convey 'truth' with as much accuracy as the knowledge and technology of the day would allow.

Nature-printing was an illustrative technique to which some authors turned as a way of enabling nature to depict itself. First explained and illustrated by Leonardo da Vinci in the late fifteenth century, it is a process by which actual leaves and other objects, are used as the ink

10 Another interesting crossover between the world of didactic, natural history model making and the world of fine art is with a comparison between the Blaschka family's glass models of marine invertebrates and the abstract glass creations of the American glass sculptor Dale Chihuly.

blocks to produce unique prints.[11] Because nature-printing is so labour-intensive and tends to destroy its subjects even as they are being recorded, it is among the rarest of the reproductive techniques.

Nature-printing reached its zenith in both quality and popularity in a sumptuous publication entitled *The Ferns of Great Britain and Ireland* (1855; p.195), written by Thomas Moore and edited by John Lindley. The book incorporated a process, patented by printer Henry Bradbury, in which actual fern specimens were pressed into soft metal to create moulds from which harder metal printing plates were made. These plates were then used to create exquisitely detailed, offset prints of the life-sized fern fronds that had been used to create them. While the ink colours employed in the published plates may not match the actual pigments found in nature, the delicate structure of the leaves are replicated more accurately in these prints than they could have been by any artist.[12]

Photography, like nature-printing, gives the illusion of capturing 'truth' without human interference, but it too requires the involvement of an artist. William Henry Fox Talbot, the British photographer who pioneered the use of paper negatives in the 1830s, made a series of what he called 'photogenic drawings', in which he recorded the leaves on light-sensitive paper without the use of a camera.[13] He referred to these leaf photographs as 'specimens', thus reinforcing the interchangeable nature of the pictures and the leaves themselves. While Talbot's cameraless pictures and the similar but better known 'photograms' of ferns and algae made by the English artist Anna Atkins in the 1840s, required less human intervention than traditional, camera-generated photographs, they nevertheless involved the active selection and manipulation of the specimens portrayed (p.162).[14] In recognition of the technical skill and artistic interpretation involved in their creation, the plant images made by Talbot and Atkins are now more highly valued as fine art than as accurate records of individual plant species as originally intended. Thus the crossover from science to art is embodied in their work in much the way it is in the more anonymous objects in Loudon's collection.

11 Da Vinci first described the techniques of nature printing in the *Codex Atlanticus* (1478–1519).

12 For more on nature-printing, see Roderick Cave, *Impressions of Nature*, British Library, London, 2010.

13 See Carol Armstrong, 'Cameraless: From Natural Illustrations and Nature Prints to Manual and Photogenic Drawings and Other Botanographs', in Carol Armstrong and Catherine de Zegher (eds), *Ocean Flowers: Impressions From Nature*, Princeton University Press, Princeton, NJ, 2004, pp.86–179. See also Larry Schaaf, *The Photographic Art of William Henry Fox Talbot*, Princeton University Press, Princeton, NJ, 2000.

14 The distinctive blue paper prints that Atkins created are called cyanotypes.

The study of taxonomy, which focuses largely on the organisation and classification of nature, is reflected in most of the models and illustrations in this collection. It is somewhat surprising, therefore, that in forming his collection and in producing this book, Loudon has made the conscious decision to refrain from structuring its contents in any predictable way. His is a collection shaped by serendipity and a discerning eye for quality and visual appeal. While many of the illustrations and artefacts that he owns have become rare and valuable through the winnowing of time, they were never conceived or created to become precious treasures beyond the reach of anyone eager to learn about the advances of science. In that sense they are democratic objects and he believes they should remain so.

Loudon is eager to share the objects in his collection with others who may not know or care about their history and prescribed purpose. He wants all who see them to allow their impressions to be shaped by their own curiosity and visual delight, not by what the academic world tells them they should see. Looking at these remarkable objects with fresh eyes, we are able to appreciate them, not just for what they were once intended to represent, but in ways their creators might never have imagined.

The Collection

by George Loudon

Instituteur Pitoiset
12 framed didactic displays
c.1890
Each: 40 × 33 × 4 cm
(15¾ × 13 × 1⅝ in)
France

The *Leçons de Choses* (Object Lessons) series was produced by a Monsieur Pitoiset for the purpose of instructing primary school children in a variety of subjects. They combine text, drawings, prints and actual objects.

'Object lessons' were introduced into French teaching in the second half of the nineteenth century. The technique involved using real objects, rather than words alone, to teach small children, and was applied particularly to science subjects, such as natural history and geography, but also to the study of craft and industry.

These boxes are one of the first objects that I acquired for my collection and in many ways they represent the essence of what fascinates me about the items that comprise the collection and that I describe in this book, namely didactic material, interestingly made, whose original purpose has been lost and whose current function is, as a consequence, open to multiple interpretations.

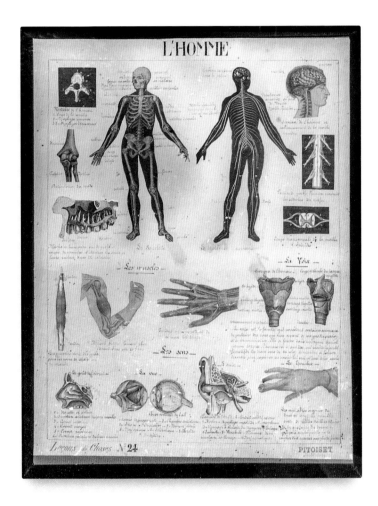

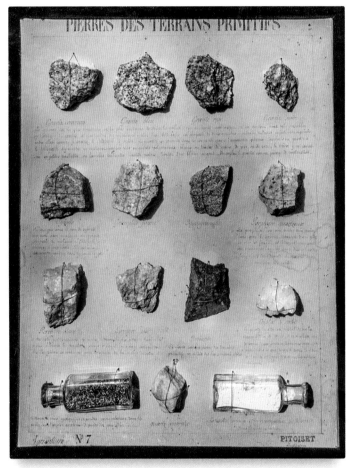

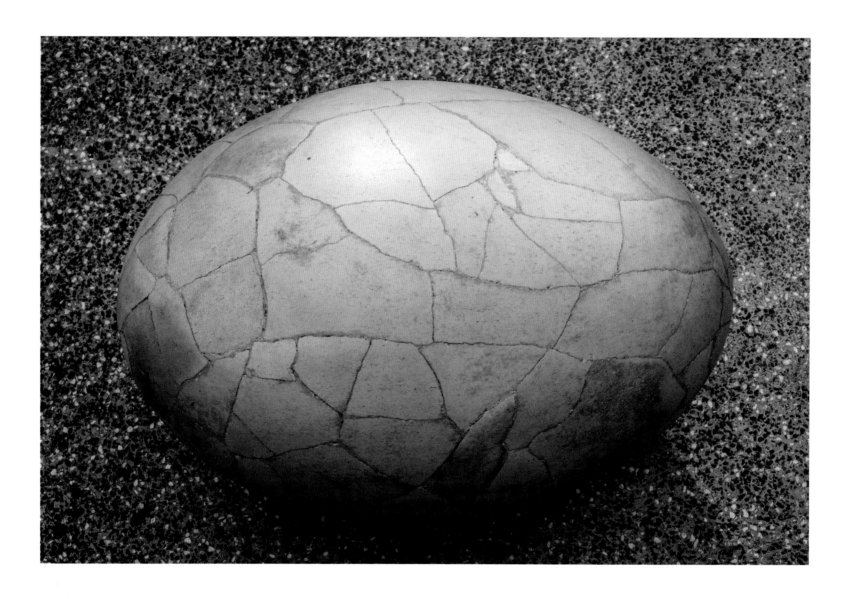

Two reconstituted *Aepyornis maximus* (Elephant bird) eggs
18th century or earlier
34 × 23 cm (13⅜ × 9⅛ in)
Not illustrated:
32 × 22 cm (12⅝ × 8⅝ in)
Madagascar

Aepyornis Maximus is the extinct Elephant bird from Madagascar. It was the largest bird ever to have lived and weighed up to 400 kilogrammes. Their eggshells are still being found, despite their extinction hundreds of years ago, which suggests that they must have been quite numerous.

Though natural history museums usually have complete eggs, I liked the fact that someone had assembled these two. It must have been quite a puzzle. Their size is, of course, amazing.

Recent DNA analysis suggests Elephant birds evolved from small flying birds after the separation of the continents. They were thought to be related to moas and emus, the flightless birds of Australia and New Zealand, which evolved at the time when Africa was part of the same land mass.

The form of the egg has appeared regularly in art. Its shape is, of course, deeply satisfying in sculptural terms and the size of this particular example adds to its wonder, hence its inclusion in the first part of the book (p.17).

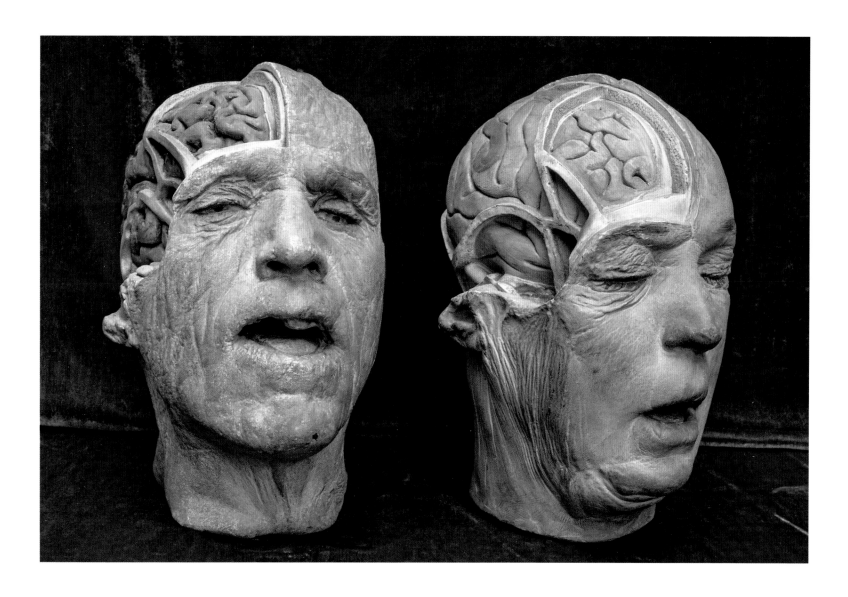

Casciani and Sons
Two plaster medical heads
19th century
Left: 25 × 17 × 22 cm
(9⅞ × 6¾ × 8⅝ in)
Right: 25 × 16 × 25 cm
(9⅞ × 6¼ × 9⅞ in)
Dublin

Casciani and Sons, figure makers and moulders based in Dublin, made these plaster heads for didactic purposes. The Cascianis came to Ireland from Tuscany in 1874 and took over the firm of one Moschini, maker of plaster statues and death masks.

These particular examples have part of their brains exposed and were modelled by Daniel John Cunningham (1850–1909), a Scottish physician and zoologist active in Dublin from 1883 to 1903, after which he moved to Edinburgh. Cunningham, like other medical men of that period, was part of the movement that lead to the idea that crime and other social ills could be explained by physical inheritance, rather than by psychological factors.

These models were used for teaching the former approach in universities such as Melbourne, while at Harvard University there is a photograph of the same heads in the Psychology Laboratory.

They are unusually ghoulish, which is why I include them in the first part of the book (pp.14–15). The reference to sculpted heads, both in classical and contemporary art, is not too hard to make.

Tehuelche people
72 Patagonian arrowheads
Date unknown
Each, between:
2–9 × 2–5 × 1–2 cm
(¾–3½ × ¾–2 × ⅜–¾ in)
Argentina

Charles Darwin's *The Voyage of the Beagle* (1839) is one of the great travel books. I love Patagonia, as did Darwin.

In calling up images of the past, I find the plains of Patagonia frequently cross before my eyes; yet these plains are pronounced by all to be most wretched and useless.

They are characterised only by negative possessions; without habitations, without water, without trees, without mountains, they support only a few dwarf plants. Why, then – and the case is not peculiar to myself – have these arid wastes taken so firm possession of my mind?

Charles Darwin, *The Voyage of the Beagle* (1839), entry for 6 August 1836.

These arrows were used both for fighting and hunting. The most common prey of the Tehuelche people would have been guanacos, mammals related to llamas, as well as rheas, the flightless bird related to the ostrich.

When, in 2011, on one of our frequent visits to Argentina, the opportunity presented itself to acquire Tehuelche arrowheads from southern Argentina, I could not, and did not, resist.

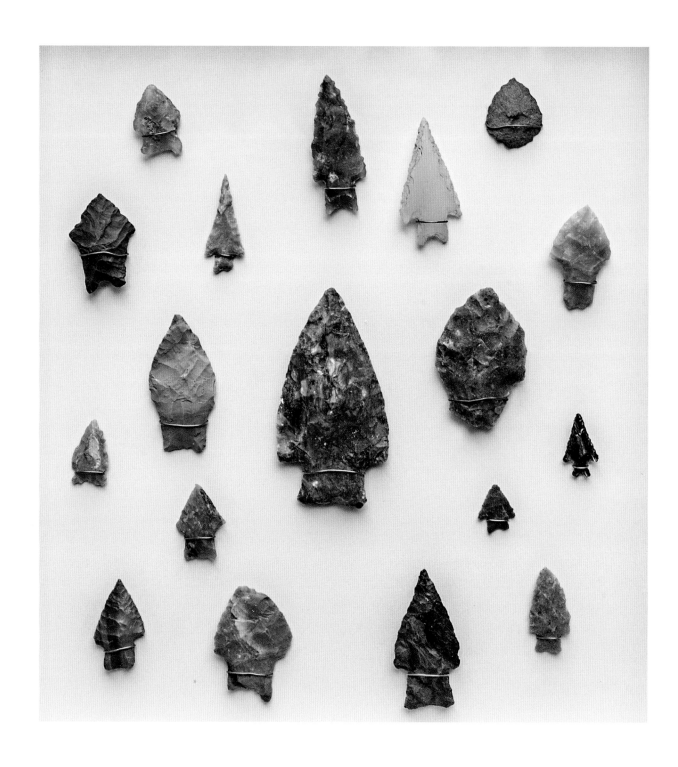

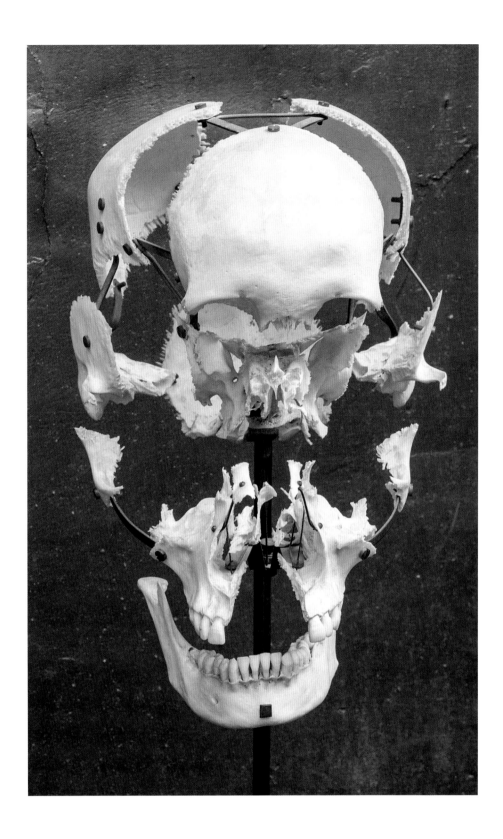

Exploded skull
c.1900
52 × 28 cm (20½ × 11⅛ in)
Europe

These exploded or disarticulated skulls are quite common in medical teaching establishments and are used to demonstrate the different bones in the human skull, a technique developed in France in the late eighteenth to early nineteenth centuries. The human skull contains 20 bones and up to 32 teeth.

This skull, used for medical teaching in the early twentieth century, once belonged to a human being. Nowadays plastic models are used in place of the real thing. In the course of the twentieth century

people began to leave their bodies to science, whereas previously only unclaimed bodies from hospitals, prisons and asylums were made available for anatomical purposes.

The process of bleaching and disassembling the constituent parts has completely de-personalised the skull of this once living person. It has become an abstract work, as can also be seen in the first part of the book (p.21).

Jones Quain, MD and
Erasmus Wilson (eds)
*A series of anatomical plates,
in lithography, with references
and physiological comments,
illustrating the structure
of the different parts of the
human body*
200 lithographic plates by
W Fairland, W Bagg, J Walsh
and WJE Wilson
1842
5 parts in 2 vols.
Height: 52 cm (20½ in)
London

This is one of the most beautifully produced set of lithographs dealing with human anatomy. It was printed in stages between 1836–42: the muscles in 1836, the vessels in 1837, the nerves in 1839, the internal organs in 1840 and the bones and ligaments in 1842. The lithographs in the work are of a particularly high quality, however very little is known about the artists who produced them.

Jones Quain (1796–1865) was an Irish anatomist who became a professor at the University of London.

He was assisted in this work by Erasmus Wilson (1809–84), a very wealthy doctor, who was knighted at the end of his life and who was well-known for his interest in skin diseases. Wilson was also interested in Egypt and personally paid for the transportation of the obelisk known as Cleopatra's Needle from Alexandria to the Thames Embankment, where the monument now stands.

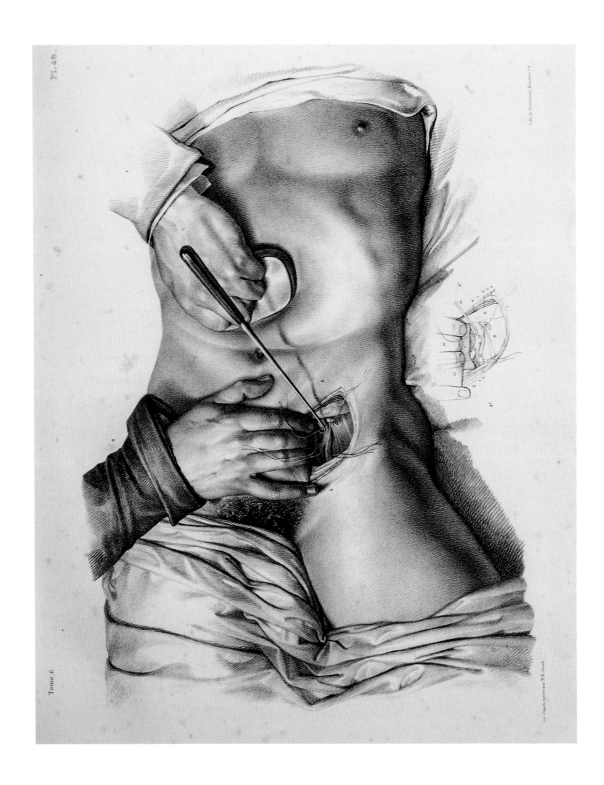

Professor Dr Ernst Haeckel
Kunstformen der Natur
1899–1904
100 plates
Height: 40 cm (15¾ in)
Leipzig and Vienna

Ernst Haeckel (1834–1919) was a charismatic German scientist: biologist, naturalist, physician and artist. He was Professor of Comparative Anatomy at the University of Jena in Germany, where he worked from 1862 to 1909 and promoted Darwin's work in that country, although he had slightly different ideas about evolution.

Haeckel made quite a name for himself working on invertebrates such as radiolarians, sponges and other sea creatures. In 1866 he met Darwin, biologist Thomas Henry Huxley (1825–95) and geologist Charles Lyell (1797–1875) on a trip to England. It was probably as a result of his expertise in his subject that he was asked to analyse part of the scientific results of the Challenger expedition (1872–76), which was the first major scientific investigation of the deep oceans around the world.

He was also a brilliant artist and this publication on the art forms of nature, his most famous, is reissued regularly. It was an influential source for artists working in Art Nouveau circles.

Although he was wrong about a number of evolutionary matters, he remains a very attractive character from late German Romanticism. Handsome, artistic, well-travelled and an eloquent communicator of science, he charmed many notable people, including the Egyptian Khedive, Ismail Pacha, who lent Haeckel one of his government steamers for his journey around the coral banks in the Red Sea.

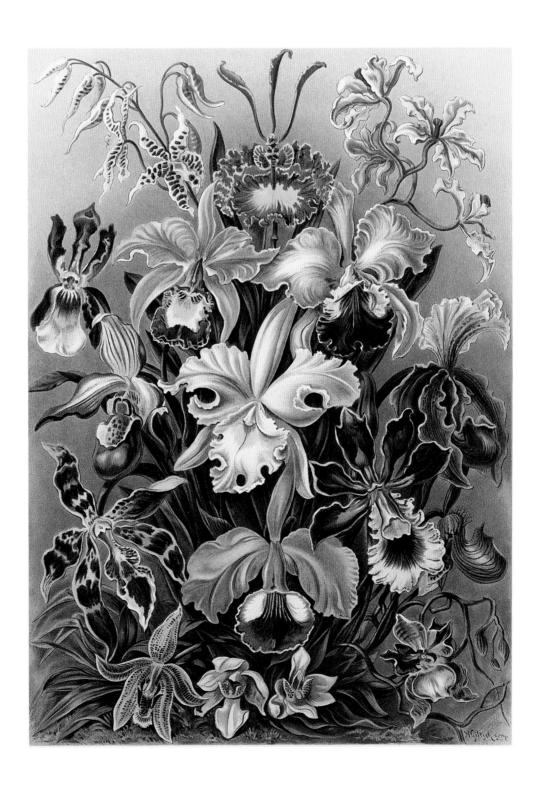

Yoichirō Hirase
Kai-Senshu
1915–22
4 vols.
Height: 25 cm (9⅞ in)
Kyoto

The collection holds a number of works that deal with the introduction of Western science to Japan. This includes botanical illustrations with dates ranging from before the Meiji restoration in 1868 up to 1922, and three early nineteenth-century globes. I find the combination of Japanese and Western science and aesthetics particularly fascinating.

Japan's opening to the West did not occur in a single rush with the Meiji restoration of 1868. The Tokugawa military regime had already started to put out careful feelers and contact with the Dutch residing since the seventeenth century on Deshima Island, in Nagasaki Bay, was intensified. Philipp Franz von Siebold (1796–1866; see p.119) and his pupil Keisuke Ito (1803–1901) were two important figures in nineteenth-century Japanese science. Ito was the preeminent Japanese botanist at the time and a professor at Tokyo University. Another botanist working in Tokyo at the time was Chikusai Kato (1813–?) who was responsible for the botanical wood panels in the collection (p.201).

One particularly interesting character was the zoologist and botanist Yoshio Tanaka (1838–1916; see pp.95, 101, 233, 242). Tanaka's brother had been a doctor on Deshima and had been in contact with the Dutch, which had enabled Yoshio to gain access to Western scientific material. He studied medicine and pharmacy under Keisuke Ito and joined him at a sort of national think tank set up in the early 1860s in response to US Commodore Perry's intervention in 1853, when he sailed into Tokyo Bay and began the process of Japan opening up to the West.

The illustrations in these four volumes dealing with 'a thousand shells' are spectacular. Yoichirō Hirase, born in 1859 and Japan's first formal 'malacologist' or studier of molluscs, founded a museum in Kyoto in 1913 using his private fortune. The museum was closed in 1919, before his death in 1925, and the majority of his valuable collection of shells was destroyed during the Second World War.

It is noteworthy that in this guide, Hirase employs full Linnaean classification, the system most commonly used in the West, alongside the equivalent Japanese system, despite the considerable trouble that earlier taxonomists had faced in reconciling these Western and Japanese classification terms.

Although these illustrations were produced in the early twentieth century, their style is quite traditional. The shell included in the first part of the book becomes very different when heavily enlarged (see pp.10–11).

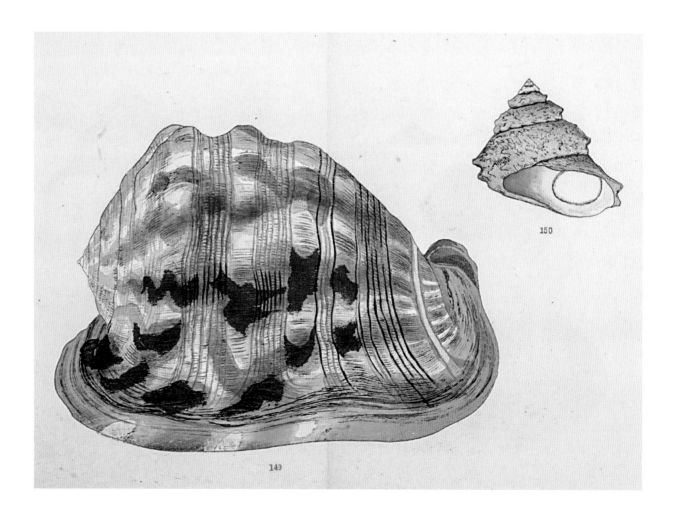

149

150

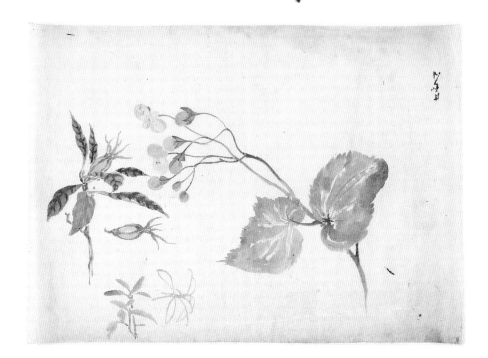

Box of botanical
watercolour paintings
1867
Each: 46 × 37 × 5.5 cm
(18⅛ × 14⅝ × 2⅛ in)
Japan

This item comprises nine botanical watercolours. Five of them are mid-nineteenth century, by an unknown artist, and reputedly commissioned by Yoshio Tanaka (1838–1916) the zoologist and botanist. In 1867, just before the Revolution leading to the Meiji restoration, Tanaka was part of the group organising Japan's participation in the International Exposition in Paris. When he returned from France, he found he was working for a new regime. He had a long and successful career in education and the promotion of all sorts of economic activity, such as useful botany (p.101) and commercial fishing (p.233).

The other four watercolours, including the two illustrated here, are likely to have been made by the Chinese artist, Saien Hōsai, whose Chinese name was Xi yuan Fang ji (1736–?). Saien was a commercial artist who had emigrated from China to Japan in 1774, before finally settling in Nagasaki. He is most famous in Japan for a painting of Mount Fuji. The date of one of these watercolours is September 1774 written in Chinese characters and following the Chinese calendar.

Kan'en Iwasaki
Honzō Zufu
1828
4 vols. with 40 loose pages
Height: 26 cm (10¼ in)
Edo (Tokyo)

This guide to Japanese flowering medicinal plants covers some 2,000 plant specimens in a total of 96 volumes, of which I have the first four and a further 40 loose sheets from volumes 8 and 89, that someone took apart.

Honzō Zufu was a very ambitious project. Published in 1828, it is a relatively early work in the collection. The hand-coloured woodblock illustrations are exquisite, although the pages are quite small.

I originally found the first four volumes in a bookshop in Tokyo, while the loose pages appeared in an auction in London quite recently.

Visiting a Tokyo antique book dealer is quite different to doing so anywhere else. There are very few old buildings in the city, so the shop is in what looks like a modern ground-floor office in an undistinguished 1960s building in a side street, with iron shelves and strip lighting. As with dealers everywhere, patience is vital. One sits on small stools, perhaps next to an English-speaking nephew of the dealer, gradually going through the offerings, knowing the best is likely to come at the end, often after an hour or two.

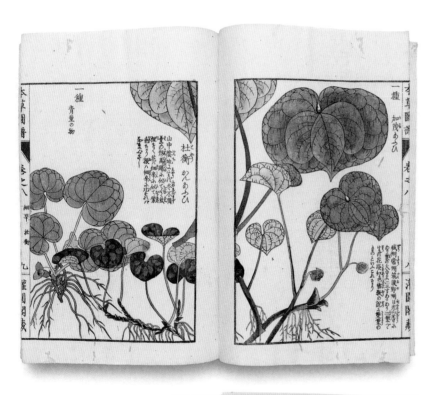

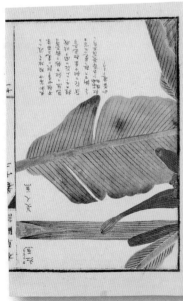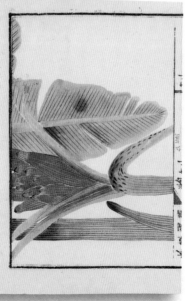

Professor Dr Ernst Haeckel
Ernst Haeckels Wanderbilder: die Naturwunder der Tropenwelt, Ceylon und Insulinde: nach eigenen Aquarellen und Ölgemälden nebst reich illustriertem text
n.d.
1 portfolio of 11 folders of loose leaves
Height: 41 cm (16⅛ in)

Ernst Haeckel was a German biologist, naturalist, philosopher and artist whose reputation as an artist was based on his natural history illustrations, such as the ones in the publication *Kunstformen der Natur* (1899–1904; p.93) as well as on his interest in painting on his travels. *Wanderbilder: die Naturwunder der Tropenwelt* (Images of Travels: The Natural Wonders of the Tropical World) resulted from a trip to Ceylon (now Sri Lanka) and the Dutch East Indies (now Indonesia).

It consists of eleven folders of unbound illustrations and text with descriptions of various locations and sites. He also somewhat randomly moves into natural history, describing subjects such as plants, jellyfish and corals. The illustrations are of his paintings, mainly landscapes, and his photographs of buildings and people, including the inevitable nudes.

What is extraordinary about this publication is that after ten folders there is unexpectedly a supplement at the end of the publication in the form of a folder titled *Apotheose des Entwickelungsgedankes von Ernst Haeckel und Gabriel Max* (The Glorification of Ernst Haeckel and Gabriel Max's Thoughts on Human Development).

Haeckel had ideas about the evolution of the human species. He qualified the human as 'Lord of the Earth', and considered the female as the summit of beauty.

For this supplement, he invited Gabriel Max (1840–1915), a very well-known painter of the time, to make the illustration shown here. Max was interested in Darwinism and spiritualism and knew everything about apes: he had a band of pet apes at home that were measured, dissected and photographed after their death. Consequently Max was well placed to show the female as the perfect '*Kunstform der Natur*', a play on the title of Haeckel's famous book *Kunstformen der Natur*, surrounded by her simian cousins.

The rest of the supplementary folder once again deals with Ceylon. In the introduction it states that this unusual publication was produced in honour of Haeckel's 50 years as a doctor, which places it in 1907.

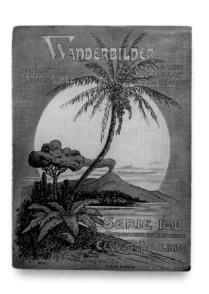

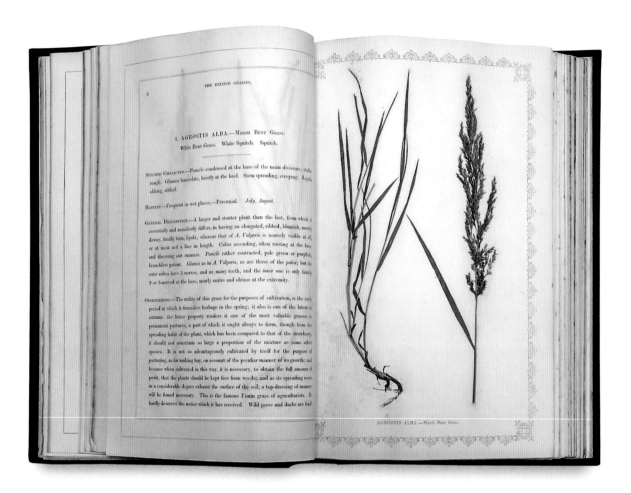

Frederick Hanham (ed)
Natural illustrations of the British Grasses
1846
62 leaves of dried specimens
Height: 32 cm (12⅝ in)
Bath, Edinburgh and London

This is a very enchanting book from 1846. It has a beautiful binding and 62 dried specimens of British grasses mounted on individual pages which the editor, Frederick Hanham calls 'natural illustrations'. Although I know nothing about Hanham, he states that his study of British grasses aims to 'produce admiration, gratitude and love to the Creator'.

The botanical descriptions of the 62 species he covers are based on the work of the botanists John Lindley (1799–1865) and William Hooker (1785–1865) – who was the first director of the Royal Botanic Gardens at Kew – among a number of others. Interestingly, Hanham adds quotations and fragments of poetry that he deems relevant to the particular plant he is describing.

Oh nature! thy minutest works amaze,
Pose close search, and lose our thoughts in craze!

Moses Brown (1738–1836)

A later publication, John Milne & Sons's *British Farmer's Plant Portfolio: Specimens of the principal British grasses, forage plants and weeds* (1890; p.239) is a more practical book and reflects the major shift in nineteenth-century agriculture in Britain from producing cereal crops to the cultivation of forage and animal husbandry.

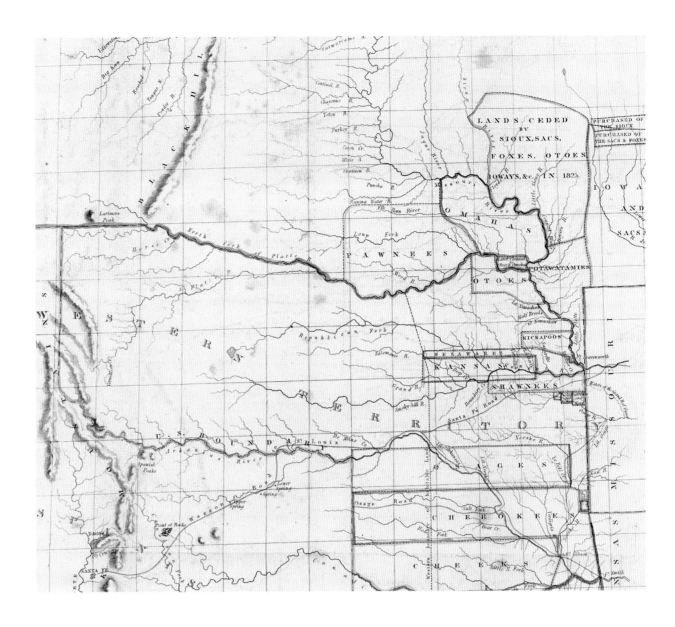

Washington Hood
Map of the Western Territory &c.
1834 (detail)
56 × 47 cm (22⅛ × 18½ in)
Washington, DC

In the 1830s the US government passed the Indian Removal Act and President Andrew Jackson (1767–1845) moved the native tribes from the Eastern States to an area west of the Mississippi, into what is now Oklahoma. This led to the infamous 'Trail of Tears', involving the tribes who refused to assimilate: the Cherokee, Muscogee, Seminole, Chickasaw and Choctaw nations.

Although the government recognised each tribe's new territory as a kind of sovereign state, white settlers moving west steadily encroached on the Native Americans' land. It is not surprising that this arrangement did not last long. Oklahoma finally became a state in 1907.

Many died on the Trail of Tears. Some of the tribes had African-American slaves, who travelled with them.

This map was created by Lieutenant Washington Hood (1808–40), whose health failed at the age of 32 during a subsequent expedition on the Mississippi.

Although this map does not have a strong aesthetic appeal and so in that sense does not fit squarely in the collection, it emanates a strong magic from the history around it and so for me qualifies. It is an incredibly sad little scrap of paper. We can speculate about a number of things: the type of character of the man that drew it; something about his life and death in the service of the US army, the miles and miles of awful terrain and weather that the tribes had to cover; and the ultimate failure of the US to provide lasting homelands for the tribes.

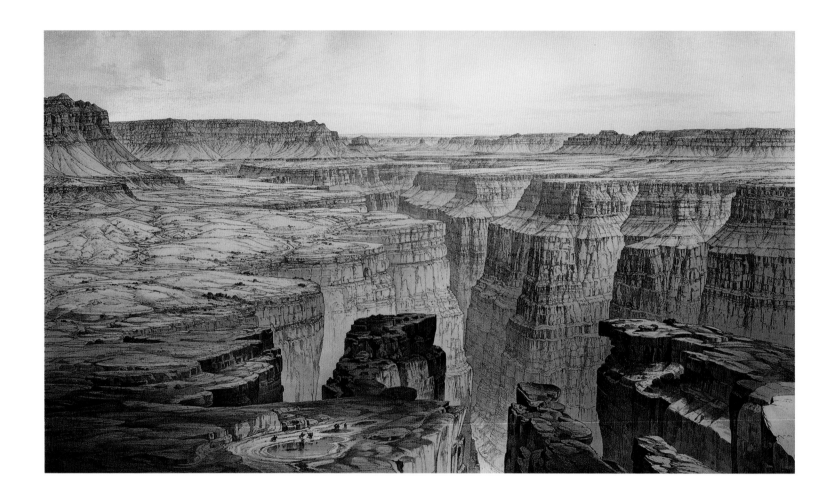

William Henry Holmes
*Grand Cañon at the Foot of the
Toroweap – Looking East*
1882
48 × 78 cm (18⅞ × 30¾ in)
Washington, DC

This is one of an amazing set of illustrations by William Henry Holmes (1846–1933) produced as part of the published work of the US Geological Survey of the Grand Canyon area. The survey was the result of an amalgamation of a number of individual state and military surveys that had taken place in the West since just before the US Civil War. All of these were part of the drive to the West after the Louisiana Purchase (1803) and the Mexican War (1846–48). Surveying these huge areas and making an inventory of the resources to be found captured the nation's imagination.

John Wesley Powell (1834–1902), the toughest and best organised of these surveyors, was appointed Head of the US Geological Survey in 1879. He attracted numerous scientists and artists to his expeditions. He also felt sympathy for the indigenous tribes, which was relatively rare at the time.

Holmes was an artist appointed to the US Geological Survey in 1872 who joined the Grand Canyon Survey in 1881. The Grand Canyon of the Colorado River is a cross section of geological time. The deepest point of the canyon contains rock that is some 2 billion years old.

The collection includes other survey maps made in 1853–55 to mark the most favourable routes for a transcontinental railroad. One of these maps shows an area marked 'unexplored South Utah and into Arizona'. Powell later became interested in this zone and in the local Ute people.

Yoshio Tanaka
Yuyo Shokubutsu Zusetsu
1891
7 vols. with illustrations by
Sessai Hattori
Height: 23 cm (9⅛ in)
Imperial Museum, Tokyo

This is an illustrated guide to useful plants – plants used in medicine, clothing, building and industry etc. – which was published in 1891 by the zoologist and botanist Yoshio Tanaka (see p.95). Tanaka was a practical civil servant interested in promoting useful botany, agriculture and other scientific endeavours for the benefit of Japan. Here he compiled seven volumes of useful plants and was officially supported by the Imperial Museum in Tokyo who published the guide.

The illustrations by Sessai Hattori (1807–?) are more practical than beautiful, and once again the Western and Japanese nomenclature is combined in a publication that is, of course, typically Japanese.

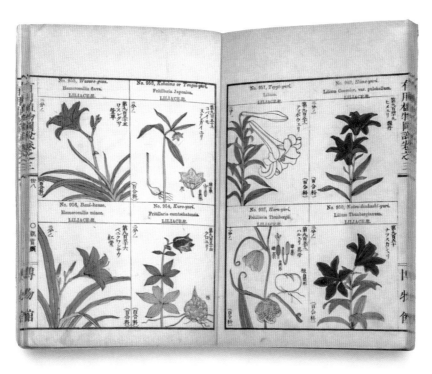

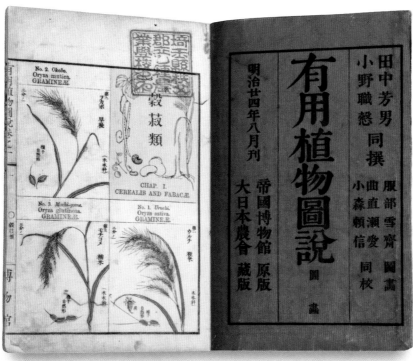

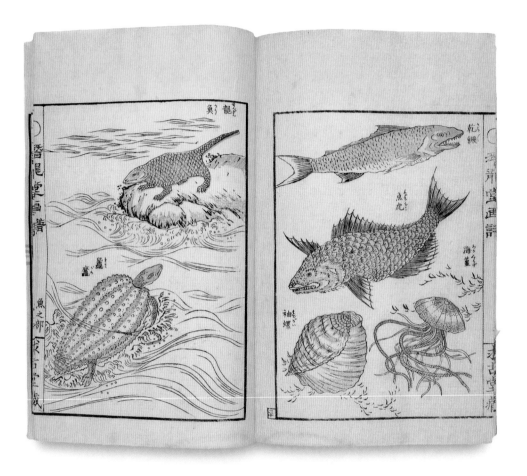

Kiyoshi Takizawa
Senryūdō Gafu Gyorui No Bu
1880
One vol.
Height: 23 cm (9⅛ in)
Tokyo

The collection includes volume one of a series of three illustrated books published and illustrated by Kiyoshi Takizawa in 1880. The first volume, *Gyorui No Bu*, covers fish and shells, while the other two cover landscape and famous people in Japan and China respectively. Though the purpose of these volumes is not clear, it seems that they were intended as simple illustrated encyclopaedias.

The woodblock illustrations here are quite elaborate. While the technique and presentation is of a typically Japanese style, the drawings themselves are much more technical and informative than the images that appear in some of the other books in my collection, which focus more on the artistic than the instructional. This volume has a certain charm.

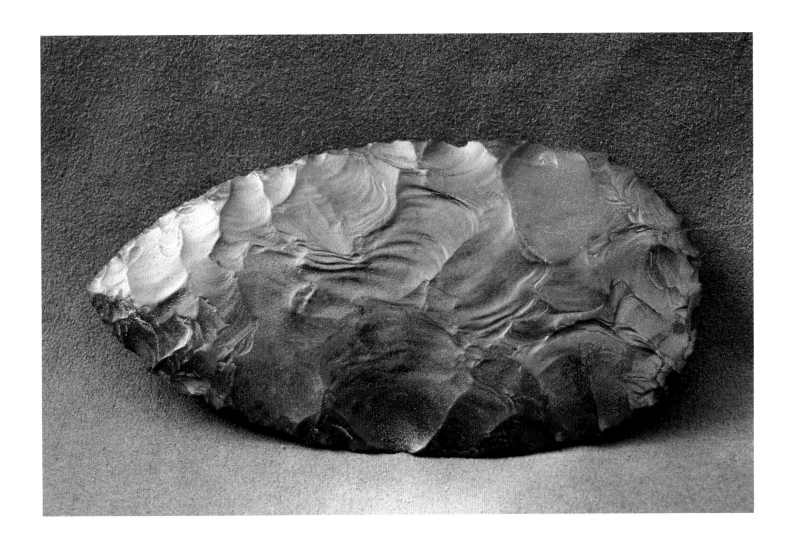

North American obsidian
hand axe
c.6000 BC
15 × 9 × 1 cm (5⅞ × 3½ × ⅜ in)
New Mexico

I bought this because I felt it was very beautiful.
It does not tell much of a story because I know
nothing about where it was found, what sort of
people made it or what was happening at the time,
however it is man-made and the result is simple
but very remarkable. As someone commented:
'you can see the rough birthmarks of its creation'.

Hand axes were first recognised as ancient stone
tools in 1797 when the Society of Antiquaries in
London was told that these objects were 'fabricated
and used by a people who had not the use of
metals… [from] a very remote period indeed, even
beyond that of the present world'. Observations
like this were the beginning of speculation about
the deep antiquity of humans and the Earth.

20 papier-mache
botanical models
1866–1927
Each, between:
26–49 cm × 11–15 cm
(10¼–19¼ × 4⅜–5⅞ in)
Germany

These models were acquired over time, but they look better as a group than individually. They were used in schools and universities, as their larger-than-life size made it possible for relatively large numbers of students to view them and study the component parts without a microscope. The papier-mache models were created from moulds. They also incorporated other materials, such as wood, cotton, pulp cane and gelatine, to add texture and detail. They were originally sold through catalogues, using dealers such as Václav Frič in Prague (1839–1916; see p.148).

Thirteen of the models we have in the collection were made by Robert Brendel (c.1821–98) and his son Reinhold (c.1861–1927), botanical model makers who were based initially in Breslau, a city that is now in Poland, and later in Grunewald, just outside Berlin.

The following models were manufactured by Robert Brendel between the 1860s and 1890s:

Buds of the common grape: *Vitis vinifera*
Flower of hemlock: *Conium maculatum*
Spikelet of winter wheat: *Triticum vulgare hibernum*
Fruit of Cypress spurge: *Euphorbia cyparissias*
Flower of meadow buttercup: *Ranunculus acris*
Spikelet of common oat: *Avena sativa*
Flower of marsh woundwort: *Stachys palustris*

While these models were made by his son Reinhold between the 1900s and 1920s:

Blossom of the domestic apple: *Pyrus malus L.* or *Malus domestica*
Leaves of common sundew: *Drosera rotundifolia*, a carnivorous plant shown eating an ant
Flower of rapeseed: *Brassica napus*
Male and female flower of the European alder: *Alnus glutinosa*
Male and female flower of the common hazel: *Corylus avellana*
A whole and a bisected flower of heart's ease: *Viola tricolour*

The other seven offer no clue as to their maker but each features in the catalogue of a distributor in Leipzig: Koehler & Volckmar. These remaining seven represent flax, the Scots pine, military orchid, white willow, marsh woundwort and two rapeseed flowers and are all from the second half of the nineteenth century.

These models have clearly inspired certain contemporary artists, which is why I included an enlarged image in the first part of the book (pp.44–45).

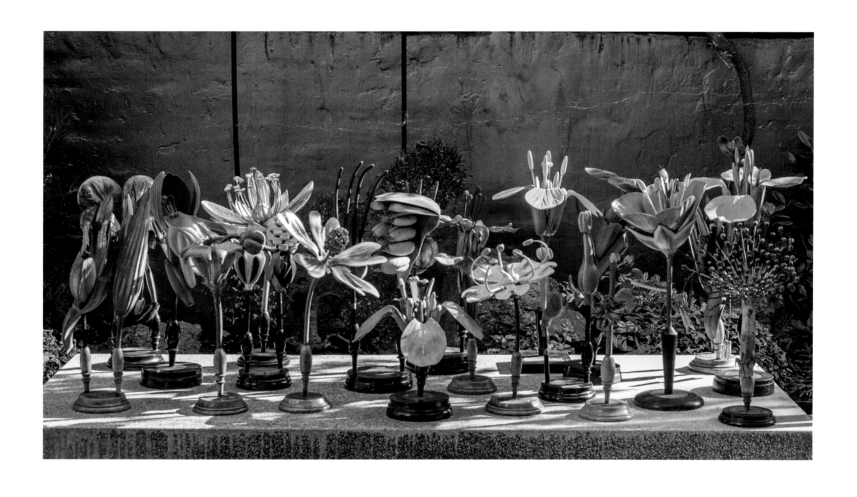

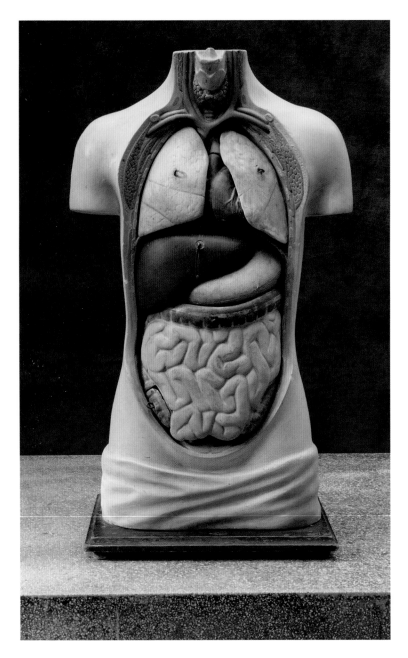
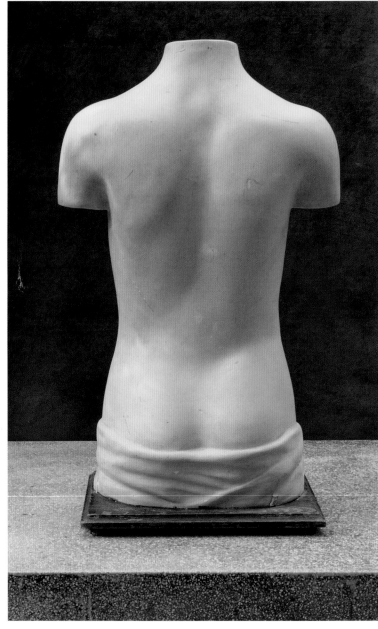

Plaster anatomical
demonstration torso
19th century
72 × 37 × 26 cm
(28⅜ × 14⅝ × 10¼ in)
France or Germany

These demonstration torsos were widely used for
educational purposes and could be taken apart in
order to study the organs. Damien Hirst used a
more modern version for his work *Hymn* (1999–2005).
He copied a children's educational toy model and
was successfully challenged for breach of copyright
by its designer and makers.

This one has a certain beauty, especially when
seen from the back. Could it be a woman? The
insides do not go down far enough to indicate the
gender of the model.

William Grosart Johnstone and
Alexander Croall
*The nature-printed British seaweeds: a
history, accompanied by figures
and dissections of the algae of the
British Isles*
1859–60
208 colour plates by
Henry Bradbury
4 vols.
Height: 25 cm (9⅞ in)
London

I have a number of nature-printed items in the
collection. This is a four-volume book of British
seaweeds, with nature-printed illustrations by
Henry Bradbury (1831–60), a controversial figure,
who was accused of stealing the latest nature-
printing techniques from the Austrians.

Nature-printing goes back a long way with early
examples dating back 40,000 years consisting of
the outline of hands painted on cave walls. In the
early days of printing, individual plants were inked
and pressed onto paper, a technique that of course
limited the number of prints that could be made
from the same specimen.

Early forms of photography used chemically
treated paper exposed to sunlight to produce an
image in shadow, a method used by such early
photographers as Anna Atkins (1799–1871, see p.162).
In the nineteenth century, various techniques were
explored to produce a metal printing plate from
an original specimen. In Austria, printer, inventor
and botanical illustrator Alois Auer (1813–69) at
the Austrian National Printing Office developed
the most successful process, making moulds of
specimens with lead or gutta-percha (a tough
plastic substance derived from certain trees), which
were then converted into metal printing plates
(see p.238). It was this process that Henry Bradbury
was accused of plagiarising.

Nature-printing was also used to print other
almost-flat objects, such as butterflies (pp.203, 223).
Finger printing and indeed photocopying and
scanning are very current forms of nature-printing.

This work on seaweeds was not Bradbury's
earliest work and, in fact, he died soon after its
completion. The publication covers more or less
everything that was known at the time and is
illustrated with 208 very fine nature-printed
illustrations, coloured subsequently by hand. This
degree of delicacy would have been impossible
to achieve at the time with any other technique.

Some of the prints have a discernable texture
when you run your finger over the page.

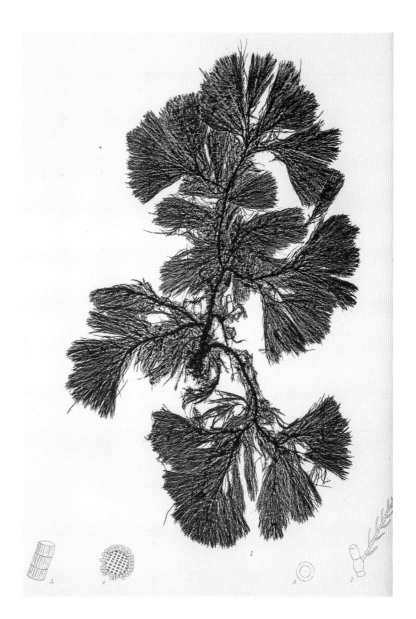

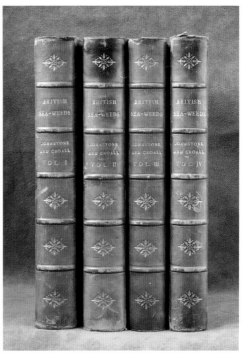

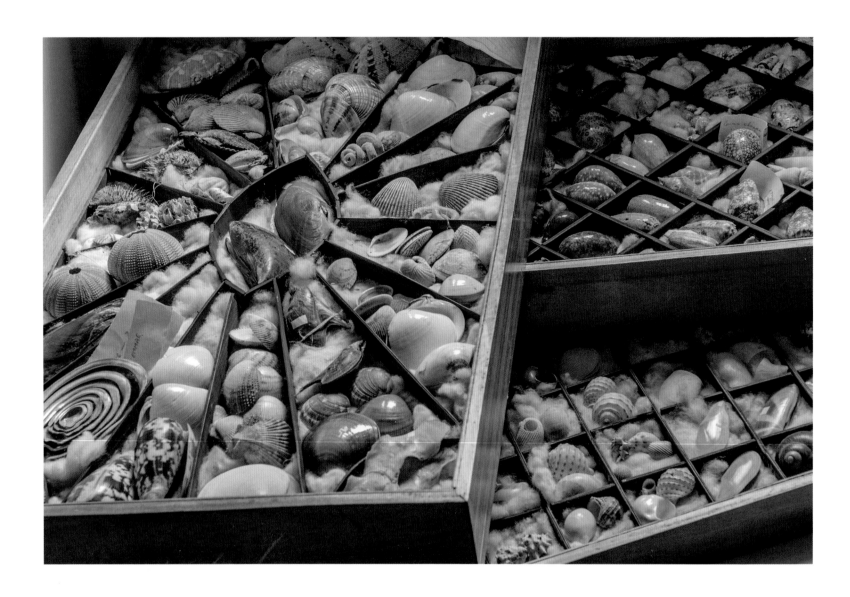

Boxed collection of seashells
Mid-19th century
72 × 41 × 46 cm
(28⅜ × 16⅛ × 18⅛ in)
Indian subcontinent

This extensive mid-nineteenth-century collection is from the waters around the Indian subcontinent. Shell collecting was already a popular activity in the seventeenth century. The Dutch East India Company began to bring back incredibly exotic shells from what is now Indonesia, leading to a 'conchylomania' comparable to the 'tulipmania' that occurred at roughly the same time in the early seventeenth century.

Eighteenth-century voyages of discovery to the Pacific – including the three that the British explorer James Cook undertook in 1768, 1772 and 1776 – further increased the European interest in shells.

The box that houses this collection is made of satinwood, inside which three removable trays hold the shells in different-shaped divisions. The whole was lovingly put together and I feel that I owe it to the original collector to have the different species identified at some stage in the future.

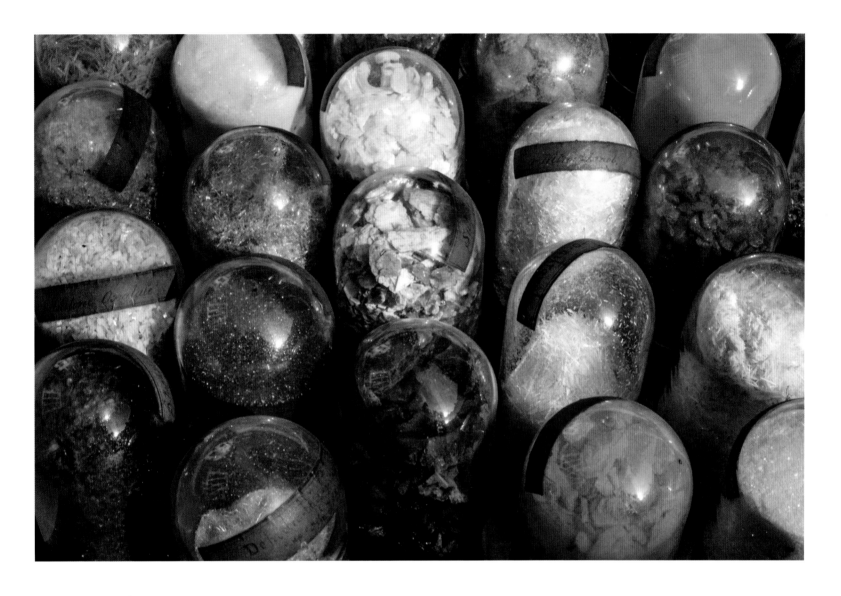

34 glass jars with dyestuffs
After 1856
Each: 12 × 5 cm (4¾ × 2 in)
Germany

In the 1840s indigo was treated with caustic potash, a process that caused an oil named aniline to emerge. In the 1850s dyes were produced from aniline on an industrial scale, particularly in Germany, where BASF (Badische Anilin-und Soda-Fabrik), one of the world's biggest chemical suppliers, was established in 1865. These dyes largely replaced earlier plant and mineral dyes.

This set of jars represents this remarkable moment of transition from the production of natural to chemical dyes in an array of colours. I imagine this set of samples was made for demonstration purposes, possibly for a school or university where the students might have learnt how to convert these chemical compounds into usable dyes.

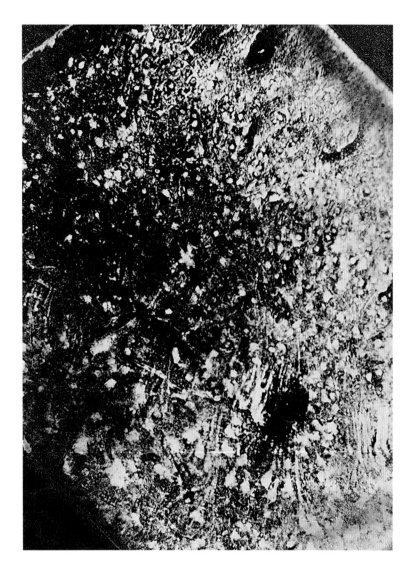
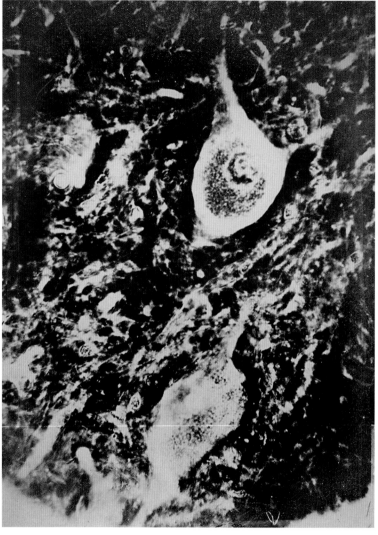

Jules Bernard Luys
Four photo-micrographs
c.1873
Each, between:
21–31 × 24–31 cm
(8¼ –12¼ × 9½–12¼ in)
Paris

Jules Bernard Luys (1828–97) was a French medic and neurologist who became well-known for his research on the anatomy and pathology of the nervous system. His first major work, *Studies on the Structure, Functions and Diseases of the Cerebro-spinal System*, was published in 1865 and was accompanied by 40 lithographic plates. At the time, Luys was criticised for the quality of the plates: critics commented that they were 'the product of his imagination'. This was always a problem when scientific observation – even early photography – was transferred by human hand onto a printing plate or stone.

In 1873, Luys published another atlas, now using unbound actual photographic plates as well as some lithographs, to demonstrate cross sections of the brain. Photography was a major step forward, particularly when it became possible in the 1870s to reproduce photographs on printing plates. The four photographs I have, of which two are illustrated, are part of the 1873 atlas. The complete body of work is very impressive and consists of 70 photographs and 64 plates.

Maison Lépine
Set of instruments for
dissection
c.1920
20 × 38 × 6 cm
(7⅞ × 15 × 2⅜ in)
France

This set of instruments does not fit squarely in the
collection, in that it has no didactic or scientific
message, although it does have a certain crude
aesthetic. The individual instruments have a beauty
that reminds me of my first collection aged ten,
of carpentry tools, which I neatly laid out, to be
looked at, never used.

   The Lépine dynasty started making cutlery in
the Lyon area of France in the eighteenth century,
a time when master cutlers provided the tools
for early surgeons. In 1860 surgical instruments
became the main focus for the Maison Lépine.
After the Second World War, the emphasis moved
to orthopaedic equipment and in 1988, after eight
generations, Philippe and George Lépine left the
company, although it kept the family name and
continues to supply surgeons with equipment
and prosthetics.

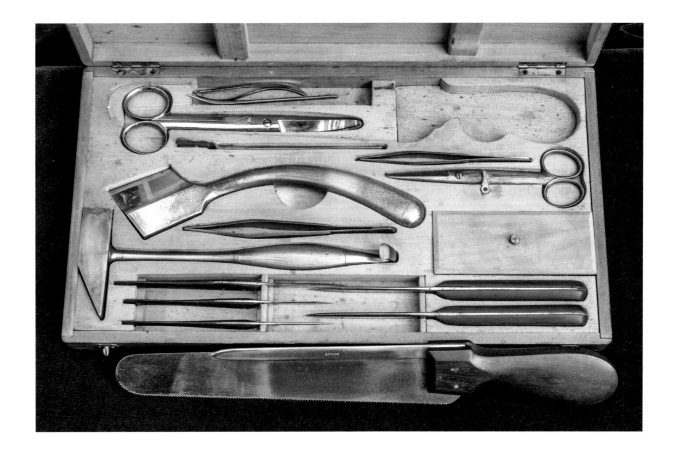

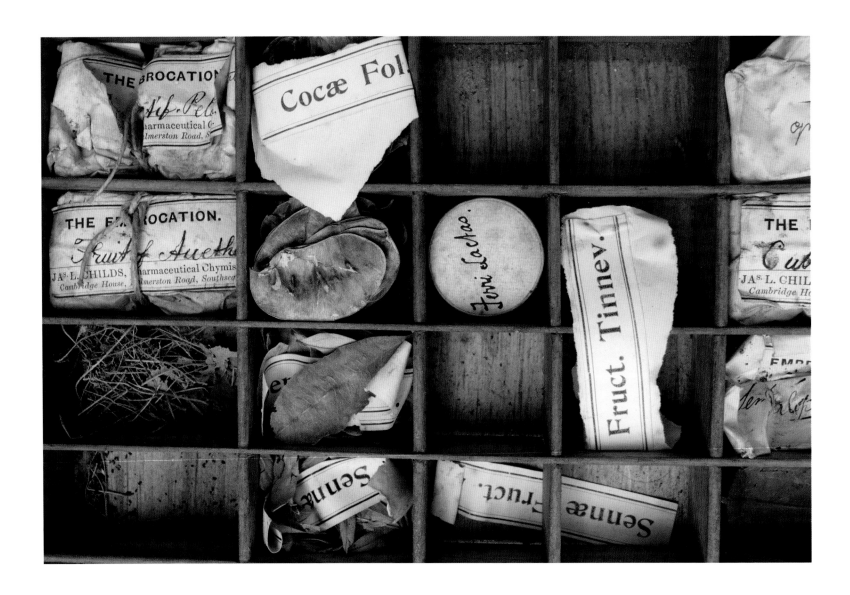

Apothecary's cabinet
19th century
49 × 33 × 48 cm
(19¼ × 13 × 18⅞ in)
England

This item is a cabinet with two outward-opening doors and nine drawers, which I saw at an auction in 2002. The drawers were jammed but the smell when you opened the doors was amazing, so I bought it sight-unseen.

The contents of the cabinet were later exposed when work was done to open the drawers and make some minor repairs. The work revealed nine layers of *materia medica*, the raw material for making medicines, which we would now call pharmacological substances. Most of the material here is botanical or herbal and kept in small glass bottles while some are larger seeds, nuts or clumps of minerals in paper packages. Most drawers contain a printed list detailing the individual products.

Six hundred and fifty materials are listed and include substances such as opium, *digitalis*, belladonna and hemlock in various forms, as well as *spermaceti*, the wax found in cavities in the head of the sperm whale and used in ointments. Also listed are benzoin, a resin from the Styrax tree, used as an ingredient of incense and believed to have a calming effect on the mind, petrified ox bile used to counter liver and gall bladder deficiencies and *pilula hydrargyri*, a medicine containing mercury which was used as a blood tonic and a cure for syphilis. Another substance, identified as *pepsina porci*, was used to stimulate the appetite. Its accompanying label reads 'gastric juice obtained from the stomach of the hog, killed fasting'.

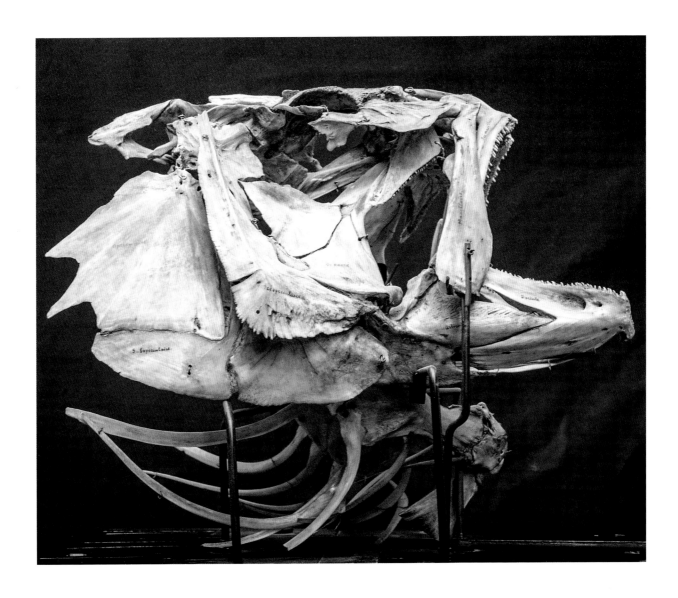

Sectionalised cod skull
Mid-19th century
37 × 45 × 43 cm
(14 5/8 × 17 3/4 × 16 7/8 in)
Europe

This is a teaching model of the head of an Atlantic cod dating from the mid-nineteenth century and would have been used for teaching the bone structure of the fish. What is unusual about this example is its size. Cod can grow up to two metres, weigh some 96 kilogrammes and live up to 25 years, however due to overfishing this rarely happens nowadays.

The different bones in the skull are labelled to help identify them and the entire structure is held together with fine wire. Originally the bones were assembled in such a way that they could be pulled out in sections, however the model is now static.

Album of insect specimens
1809
32 × 24 × 4 cm
(12⅝ × 9½ × 1⅝ in)
England

This is a very charming album of some 215 insect specimens collected in the early nineteenth century, mounted under sheets of mica and framed with printed decorative paper. They are all identified both by their Linnaean Latin nomenclature and by their common English names, handwritten on labels in ink.

The only locations mentioned regularly throughout the album are Rougham and Suffolk, which one would assume refers to the Rougham just outside Bury St Edmunds in Suffolk, as opposed to the one in Norfolk.

One or two of the insects are accompanied by personal anecdotes, for example:

> *Hippobosca avicularia*
> This is the first insect I ever collected.
> I was struck at finding it among the
> feathers of a jay shot Aug, 1809.
> It probably lives on minute acari &c.
> or would never be suffered by the bird.

This species is now known as *Ornithomya avicularia*, sometimes referred to by the common name of bird louse fly. It is a biting fly in the *Hippoboscidae* family of louse flies, all of which are parasites of birds, which means that the collector was wrong as, in fact, the fly was feeding off the bird, rather than on other parasites.

This album was in the collection of Graham Pollard (1903–76), a British bookseller and bibliographer who, interestingly, exposed the book collector Thomas J Wise as a fraud. While Wise was viewed as an eminent scholar of the late Victorian world, he was at the same time a very serious forger of pamphlets ostensibly by great English poets including Matthew Arnold; Elizabeth Barrett Browning; George Eliot; John Ruskin; Percy Bysshe Shelley; Alfred, Lord Tennyson and William Thackeray. It is said that Wise razored out pages of Shakespeare's First Folio from the British Library in order to complete copies that he later sold.

I bought this in Los Angeles from a Hollywood book dealer, which is not exactly where you would expect to find this sort of book, but perhaps film stars do collect interesting things.

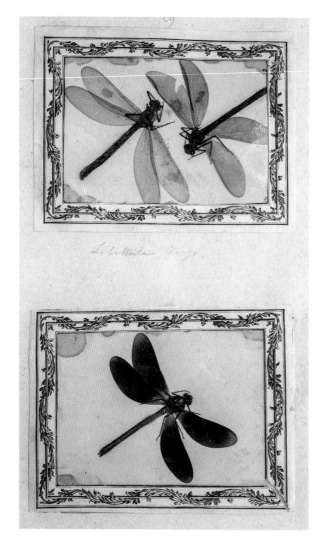

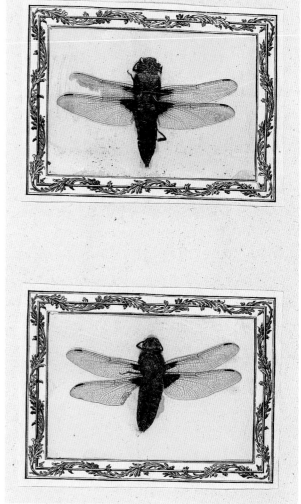

H Noël Humphreys
The genera and species of British butterflies: described and arranged according to the system now adopted in the British Museum: illustrated by plates in which all the species and varieties are represented, accompanied by their respective caterpillars, and the plants on which they feed
1859
32 leaves of colour plates
Height: 27 cm (10⅝ in)
London

The genera of British moths: popularly described and arranged according to the system now adopted in the British Museum: illustrated by a series of picturesque plates, exhibiting the insects in their different ages, with the caterpillars and the plants on which they are generally found
1860
62 leaves of colour plates
Height: 27 cm (10⅝ in)
London

Henry Noël Humphreys (1810–79) was a naturalist, entomologist and a skilled illustrator in his own right. His books of butterflies and moths were beautifully produced.

The butterfly book has very fine illustrations of the animals and the plants on which they are found. Humphreys used colour lithography – a process that was introduced in the late 1830s and involved the use of a separate lithographic stone for each colour – rather than producing a single-colour lithographic print that was subsequently hand-coloured.

Humphreys's other book, on moths, is similarly beautifully produced with 62 leaves of colour plates. Two different publishers, namely Allman and Gerrard, produced editions of this book around 1860. I have found no explanation for why these identical editions should have had different publishers.

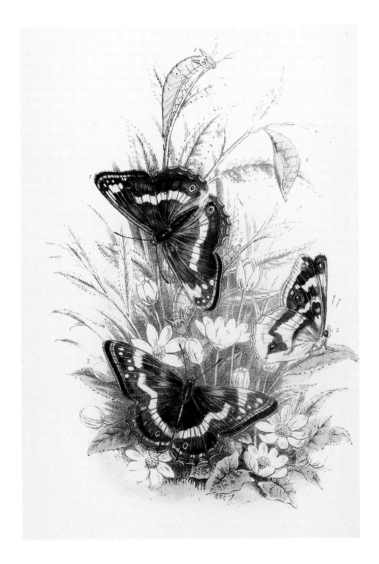

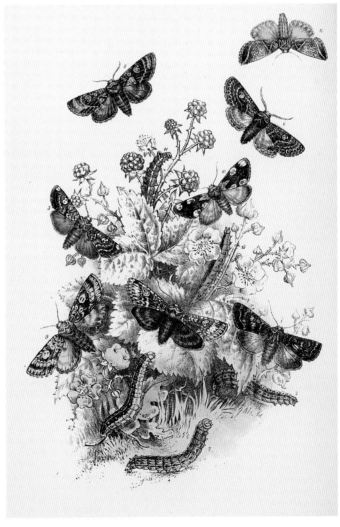

Model of the bones
in the foot
Late 19th century
37 × 66 × 32 cm
(14⅝ × 26 × 12⅝ in)
Europe

This object is very clearly a teaching model. Its wildly exaggerated size would have allowed many students to see it from a distance.

The model shows quite extensive wear and tear, so it must have been used a great deal. There are six faint signatures on various bones from what look like German students and it is just possible to make out four names: Jurgen Bremer, Dietmar Kraft, Thomas Kraus and Frank Mueller. Nothing is known about what happened to these men subsequently. Why would they do this? Medical students are well known as jokers, so perhaps this was a form of medical graffiti.

Somebody has also painted the toenails black which further adds to the mystery surrounding the model – a curious object.

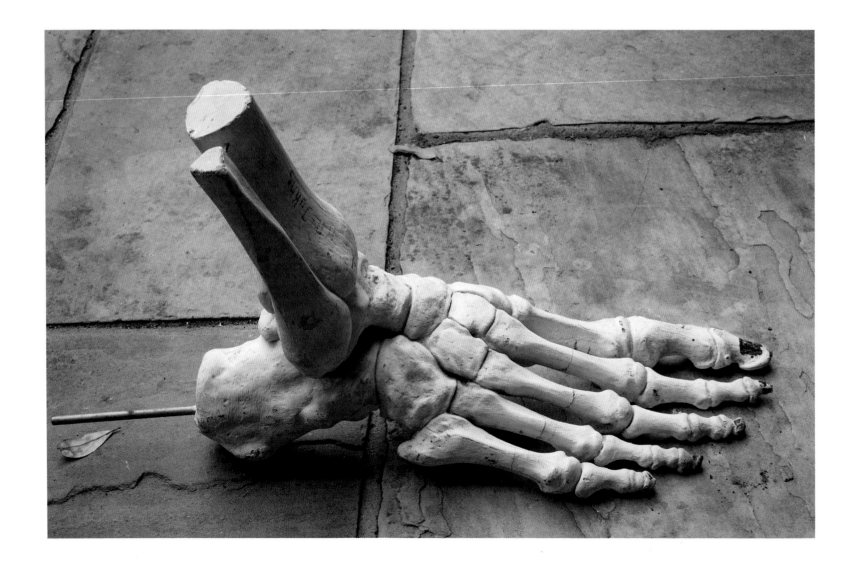

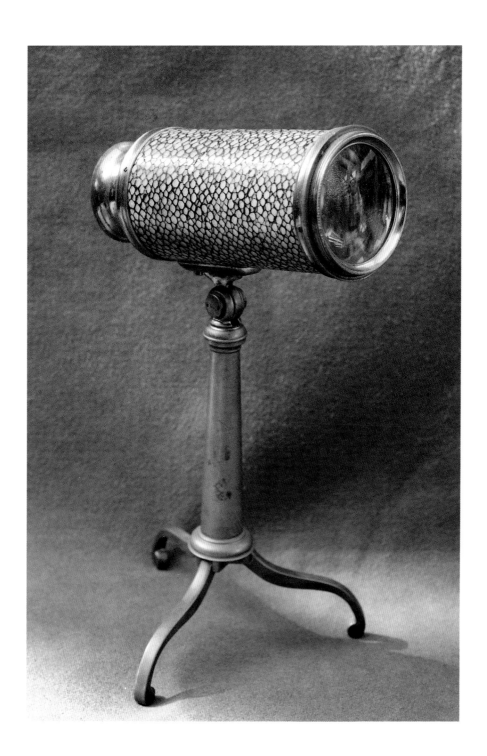

John Dollond
Tabletop telescope
Mid-18th century
19 × 5 cm (7½ × 2 in)
London

Scientific instruments do not fall squarely into the collection, as they have no didactic or scientific content. Nevertheless I found this object so beautiful, with its shagreen (leather from shark or rayfish skin) decoration. It reflects the eighteenth-century desire to combine scientific purpose with beauty.

John Dollond (1707–61) was the son of Huguenot refugees in London who made optical and astronomical instruments and was appointed optician to King George III in 1761. This particular instrument would have been made for a wealthy amateur. Dollond's optical business – Dollond and Aitchison – survived until 2009 when it merged into Boots Opticians.

Dr Louis Thomas Jérôme
Auzoux
Papier-mache model
of a wheat grain
c.1910
12 × 23 × 14 cm
(4¾ × 9⅛ × 5½ in)
France

I have ten models made by Dr Louis Thomas Jérôme Auzoux (1797–1880), including models of human anatomy (p.151), mushrooms (p.146), animals (pp.143, 157, 178) as well as this signed botanical model, produced in 1910. It depicts a wheat grain and comes apart so that students can learn how the grain is constructed.

Auzoux was a French doctor who in the 1820s started to make accurate human and animal anatomical models. By this time there was a considerable shortage of cadavers on which the increasing numbers of medical students could practice. Auzoux's models helped to fill this gap effectively as they were accurate and could be taken apart easily, an advantage over other similar wax models that were also more expensive. Like their wax counterparts, Auzoux's models were made using casts, which were lined with papier-mache, painted and numbered.

Auzoux's variety of output was enormous. By the First World War the firm was making some 320 different models, of which more than a hundred were of human anatomy. The models were widely sold within Europe and the United States. Auzoux's business continued until well into the late twentieth century, by which time it was using more modern resins.

The Boerhaave Museum in Leiden is a centre of knowledge on these models and their conservation.

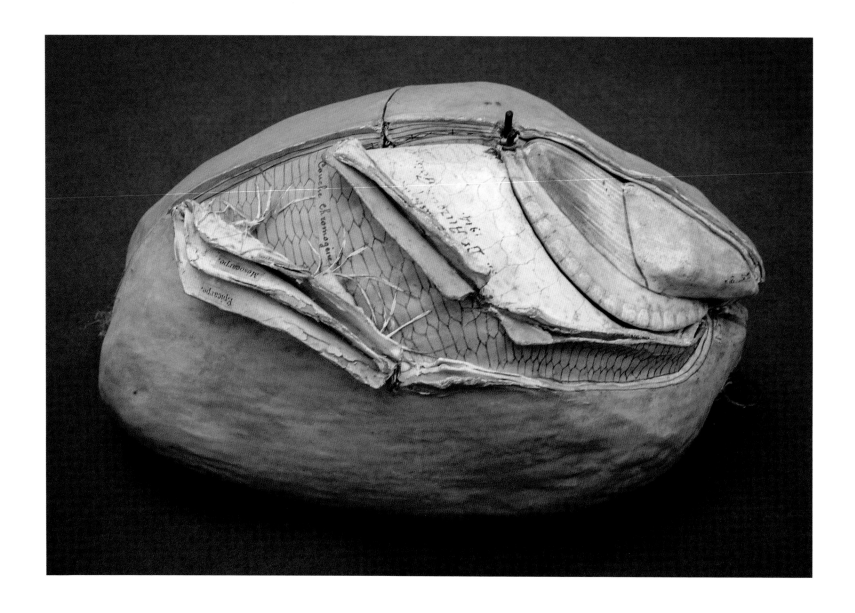

Philipp Franz von Siebold
Fauna Japonica:
Reptilia elaborantibus
1838
27 leaves of plates
Height: 39 cm (15⅜ in)
Leiden

Philipp Franz von Siebold (1796–1866) was a prolific naturalist and collector who was based in Japan in the first half of the nineteenth century. He was German, studied medicine and was recruited by the Dutch as a surgeon in the Dutch East Indies. In 1823 he was sent to Deshima, an artificial island in the bay of Nagasaki in Japan, and the only place where foreigners – in this case the Dutch – were allowed to reside. From 1823 to 1829 he made himself popular locally and was able to operate with a degree of freedom. He made a substantial collection of plants, seeds, animals and everyday Japanese tools.

In 1829 however he was expelled from Japan when he was discovered to be in possession of several detailed maps, which the Japanese viewed as tantamount to spying. Following his expulsion he settled in Leiden in The Netherlands and brought with him the more than 10,000 items that he had collected.

In 1858 he was allowed back to Japan, this time working as an advisor for the Dutch trading company Nederlandsche Handel-Maatschappij, but his usefulness had come to an end. The country was beginning to open up to Western influence and the Dutch were not the only ones the Japanese were listening to.

He returned to Europe in 1862 with a collection of even greater significance than his first. It was subsequently bought by the British Museum in London.

Von Siebold's *Fauna Japonica*, which deals with Japanese reptiles, is a typical example of Western and Japanese science mingling. The species are named both in Japanese and Linnaean terminology, and the illustrations are completely Western, even though they were made by a Japanese artist. The Smithsonian's database lists Kawahara Keiga (1786–c.1860) as the principal illustrator of this publication. Keiga was from Nagasaki and in 1811 was designated 'picture-maker allowed to work in Deshima', i.e. for foreigners. He started to act as household artist for von Siebold, accompanying him on an official mission to Edo (now Tokyo) in 1826. When von Siebold got into trouble in 1829, Kawahara did as well.

In the first part of the book I have included the most spectacular reptile illustrated in this book, the giant salamander, which can reach a length of 140 centimetres or nearly 60 inches (pp.28–29).

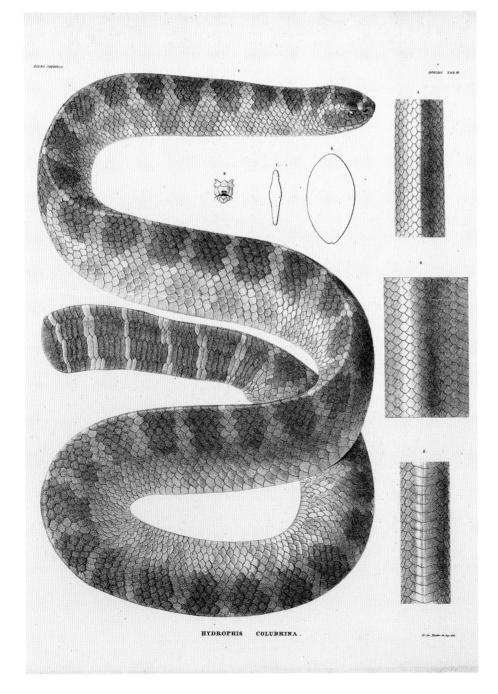

HYDROPHIS COLUBRINA.

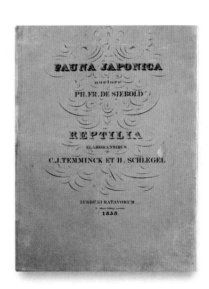

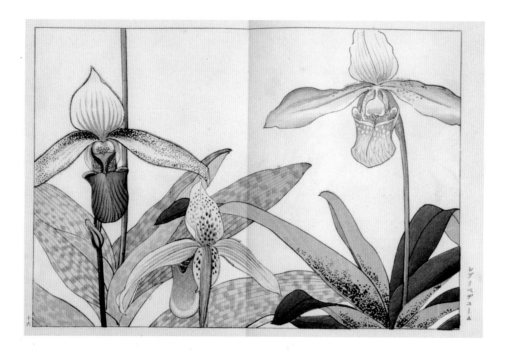

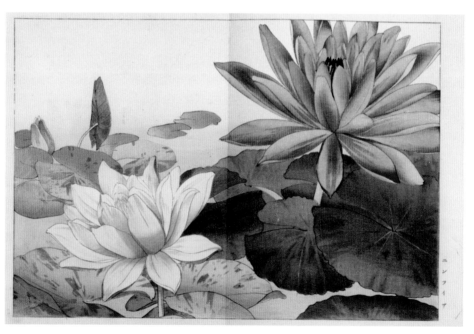

Kōnan Tanigami
Seiyō Kusabana Zufu
1916
5 vols.
Height: 28 cm (11⅛ in)
Kyoto

This publication from 1916 is quite late in relation
to other Japanese books in the collection. The book
is an illustrated guide to Western flowers, a subject
that was highly fashionable at that time in Japan.

The illustrations are really quite spectacular and
the influence of Western Art Nouveau is evident.
They exaggerate the texture of the plants amazingly,
hence the inclusion of one in the first part of the
book (pp.30–31).

The nomenclature is both Western and Japanese.
The outcome is a very unusual publication, of which
volumes one and two cover spring flowers, volumes
three and four summer flowers and volume five
autumn and winter flowers.

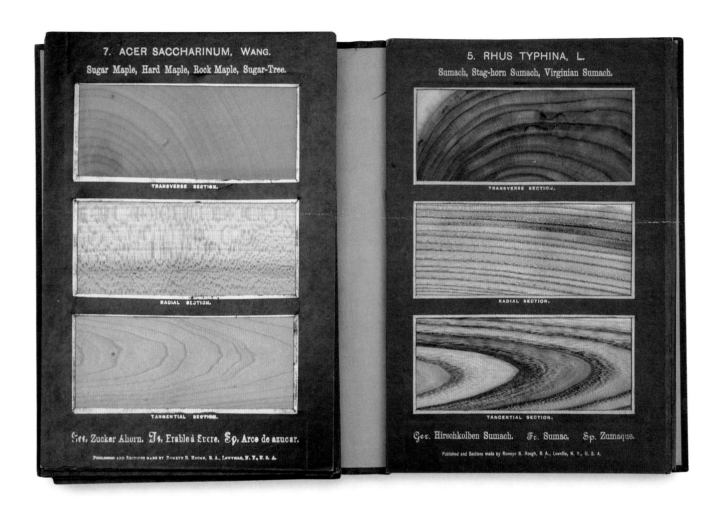

Romeyn B Hough
The American woods: exhibited by actual specimens and with copious explanatory text.
Part 1 etc.
c.1888
81 samples of wood illustrating 27 species, mounted in 354 card mounts
Height: 25 cm (9⅞ in)
Lowville

Romeyn Hough (1857–1924) is one of my heroes. He was born in Upstate New York, where his father was a professional dendrologist. Hough graduated from Cornell University, Ithaca in 1881 and did odd jobs as a naturalist for Cornell, the Smithsonian in Washington, DC and the College of Physicians and Surgeons in New York before starting on this work on American woods.

He also devised a machine for cutting very thin – from one tenth to one twelve-hundredth of an inch – sections of wood and discovered that the transverse section of some woods was very strong, and suitable for business cards and other novelties. He started a very successful business in these products alongside his scientific work.

Mine is volume one of 14 volumes that Romeyn Hough produced between 1888 and 1913, which cover all American wood species. Each volume has a text describing each species with samples of wood, wafer-thin, including transverse, radial and tangential sections, on unbound card mounts, the whole contained in a slipcase.

Some of the species he described are already extinct.

It was a monumental effort to collect all the specimens, so these volumes were produced over a number of years. It is extremely rare to find all 14 together outside specialist libraries. I recently tried to buy an earlier, virtually identical book of German woods, which used the same technique. It would, of course, have been satisfying to have acquired both books.

Richard Lawrence
An inquiry into the structure &
animal oeconomy of the horse:
comprehending the diseases to
which his limbs and feet are
subject, with proper directions
for shoeing, and pointing out a
method for ascertaining his age
until his twelfth year: to which is
added, an attempt to explain the
laws of his progressive motion,
on mechanical and anatomical
principles: the whole illustrated
by eighteen copper plates
1801
18 leaves of plates
Height: 27 cm (10⅝ in)
Birmingham

Four models of horses' legs
Mid-19th century
Legs, length, each:
34 cm (13⅜ in) (approx.)
Box: 14 × 40 × 28 cm
(5½ × 15¾ × 11⅛ in)
London

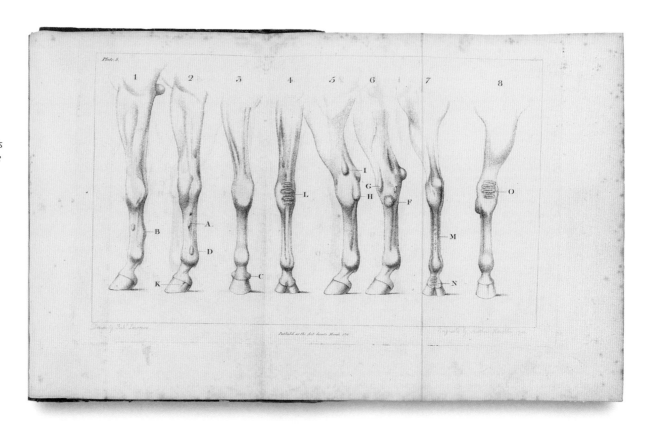

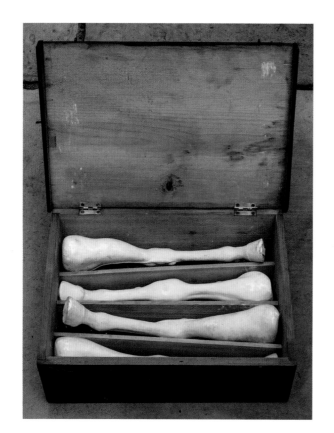

Richard Lawrence (?–c.1828) was an early vet,
graduating from the Veterinary College in London
in 1794. Subsequently, operating from Birmingham,
he wrote a number of books about horse anatomy
and pathology. He was a good artist and was
known for his paintings and sculptures of animals,
particularly horses. There is an unsubstantiated
suggestion that he may have been the brother of
the painter Thomas Lawrence (1769–1830).

In 1804 Lawrence sought permission from the
naturalist, botanist and patron of the natural
sciences Sir Joseph Banks (1743–1820) to draw the
merino sheep kept at Kew Gardens, which Banks
had smuggled from Spain to improve the quality of
British wool.

This is the first of the two books I have by
Lawrence, written 15 years earlier than *The Complete
Farrier* (1816; p.214). It is one of the earliest books in
England to deal exclusively with the horse, arguably
the most profitable area of veterinary practice.
Judging by the list of subscribers, which names 230
individuals, interest in the book was considerable.

The boxed set of plaster models, also by Lawrence,
shows the afflictions that can affect the legs of
horses, as described in the book shown here. These
models must have had some practical use, but
what that was is not clear. Did he perhaps carry
them around for demonstrations or for teaching
purposes? I found this box in Vienna.

Richard Owen
*Memoir on the pearly nautilus
(Nautilus Pompilius, Linn):
with illustrations of its external
form and internal structure*
1832
15 leaves of plates
engravings drawn by Owen
Height: 29 cm (11⅜ in)
Council of the Royal College
of Surgeons, London

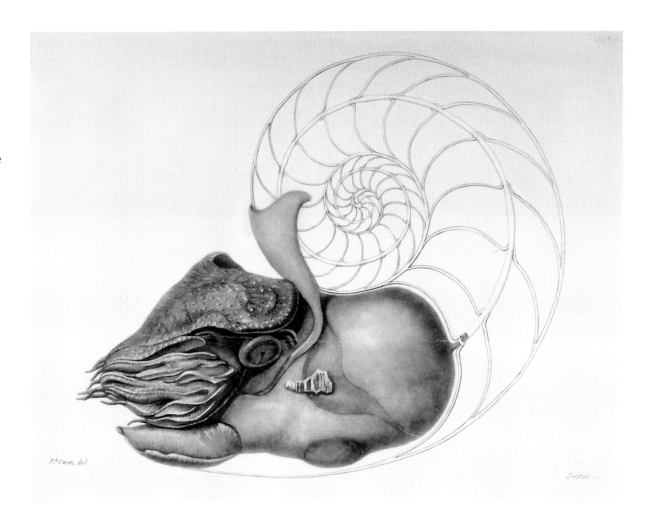

Richard Owen (1804–92) dominated natural history in England in the nineteenth century. He trained as a doctor and joined the Hunterian Museum, which is part of the Royal College of Surgeons in London. There he identified some 13,000 specimens and became an expert in comparative anatomy. He was friendly with Baron Georges Cuvier (1769–1832), the great French palaeontologist and founder of the field of comparative anatomy, who visited London from Paris in 1830.

In 1836 Owen became the first Superintendent of the British Museum's natural history departments. From there he campaigned for a new museum dedicated to natural history, which was realised in 1881 and still exists as the Natural History Museum in London.

During his lifetime he published more than 600 scientific papers. He coined the term 'dinosaur'. He also, famously, cut up the bodies of exotic animals that had died in London Zoo.

Owen was a favourite of the conservative Victorian elite and taught biology to Queen Victoria's children. He was, however, less popular with some fellow scientists, who suspected him of claiming the glory for discoveries others might have made. Owen was critical of Charles Darwin's writing and particularly of his ideas on evolution.

His *Memoir on the pearly nautilus* (1832) was produced when he was still at the Royal College of Surgeons and includes superb illustrations by Owen himself. This is once again an example of a scientist who is also a more than accomplished artist. Later in his career he used others to produce illustrations, probably due to the pressure of work.

The pearly nautilus has existed for 500 million years and has been viewed as a 'living fossil'. Its shell has always been a symbol of natural perfection. The internal chambers are connected, allowing it to adjust the gases in each chamber to control its buoyancy.

> This is the ship of pearl, which, poets feign,
> Sails the unshadowed main,-
> The venturous bark that flings
> On the sweet summer wind its purpled wings
> In gulfs enchanted, where the Siren sings,
> And coral reefs lie bare,
> Where the cold sea-maids rise to sun their
> streaming hair.

Oliver Wendell Holmes Sr, *The Chambered Nautilus* (1858).

123

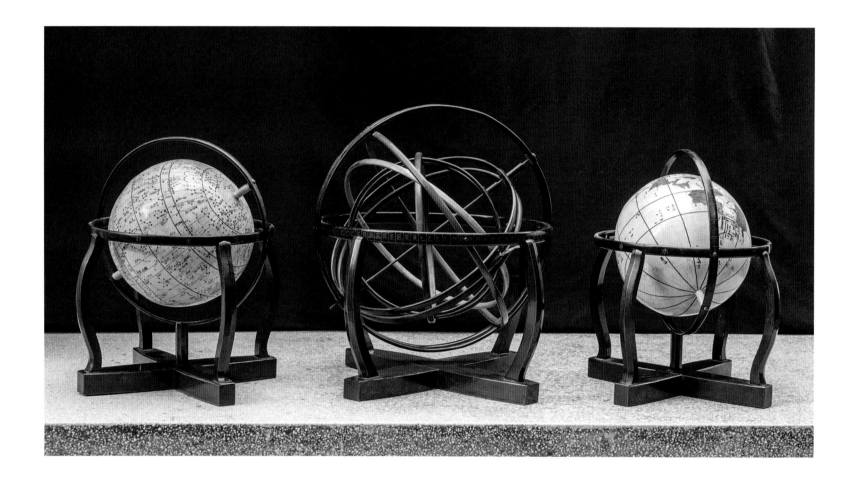

Three globes with handbook
and chart in box
Early 19th century
Left: 34 × 20 cm
(13⅜ × 7⅞ in)
Centre: 86.54 × 38 cm
(34⅛ × 15 in)
Right: 34 × 20 cm
(13⅜ × 7⅞ in)
Japan

These terrestrial and celestial globes are typical of the meeting of Western science and Japanese ideas of the world and universe. They come with an astronomical chart and handbook in Japanese, all contained in a wooden box.

Before the mid-nineteenth century Japan was cut off from regular communication with the rest of the world, except for the import of Chinese scholarly works and what was learnt from Dutch traders. The Chinese principally learnt about Western science, particularly astronomy and medicine, from the Jesuits who were successfully operating at the Chinese imperial court. However, by the late eighteenth century the Jesuits' knowledge of Western science was somewhat out-of-date and, as a consequence, the Japanese were receiving conflicting information from China and the Dutch. This was particularly true in the field of medicine, of which I unfortunately have no examples, but also of astronomy.

In these globes Western science meets Japanese artistry.

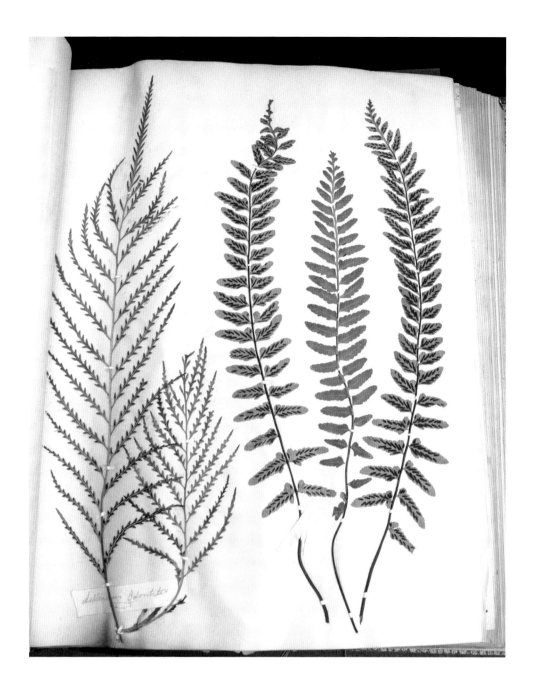

Album of fern specimens
19th century
63 × 49 × 8 cm
(24¾ × 19¼ × 3⅛ in)
England

The Victorians collected ferns to such an extent that the term 'Pteridomania' or 'Fern Craze' was coined in 1855 by the priest and author Charles Kingsley (1819–75). Indigenous ferns were diverse and abundant in the wetter parts of Britain. Somewhat later, various parts of the British Empire added to the variety of fern species, particularly from New Zealand. Fern motifs were used widely in Victorian decoration and the practice of growing ferns indoors became extremely popular.

This very large volume of dried fern specimens – mounted on sheets of paper and bound in an album – includes no printed information at all. It is known only that the spectacular size of this volume, together with the beauty of the specimens themselves, makes for a very attractive item.

Over time the bulkiness of the ferns has become imprinted into the paper on which they are mounted. If one looks carefully the pressure from the fern on the preceding and the facing pages is clearly visible (see detail, p.19).

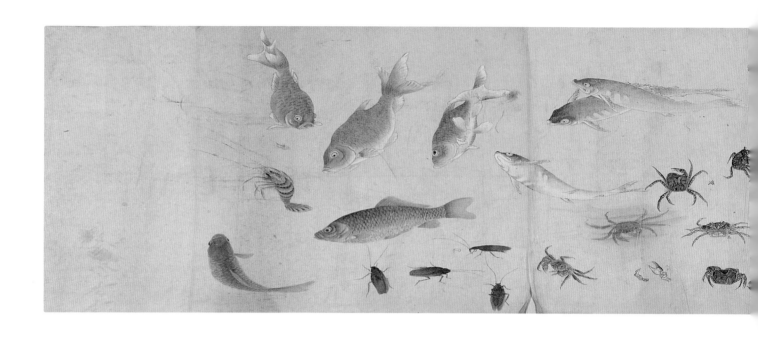

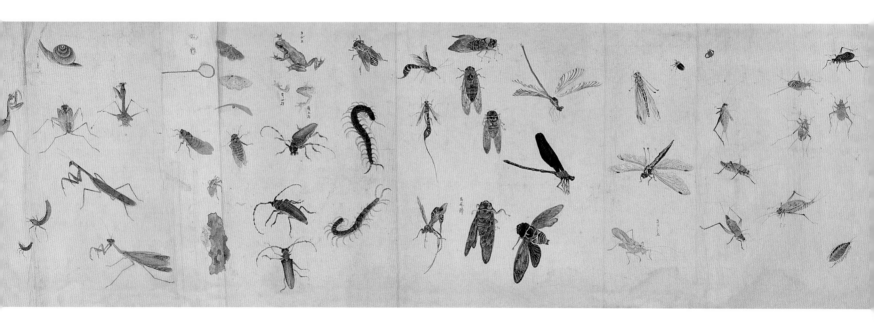

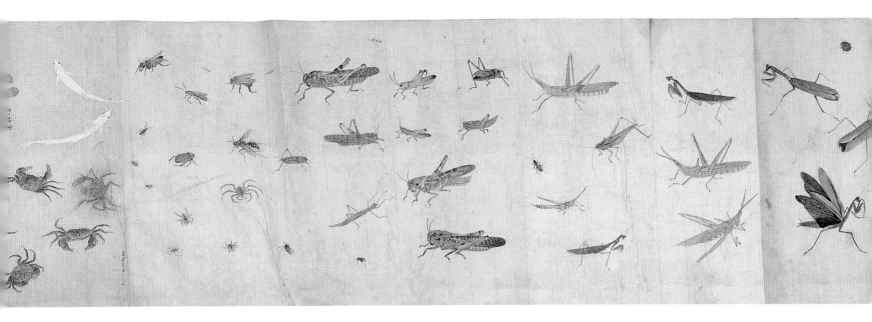

Teaching scroll with fish,
insects and amphibians
1843
27 × 262 cm (10⅝ × 103⅛ in)
Japan

In this item the artist has chosen to show all these animals in the form of a two-and-a-half-metre long scroll, rather than in book form. It is not immediately clear how this would have been used as only a few descriptions accompany the hand-painted illustrations, though this might suggest that a teacher would instruct the pupils verbally on each of the animals.

The range of animals is quite arbitrary. It covers fish, insects and amphibians but omits mammals and birds.

Although it is very difficult to see the detailed beauty of the animals in reproduction, for me this is a very fine piece of Japanese natural history art and the only Japanese scroll in the collection.

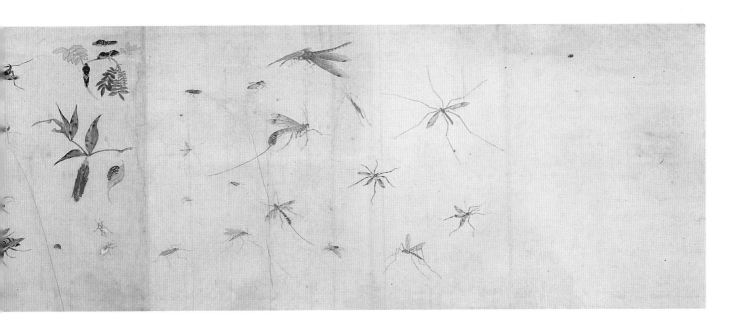

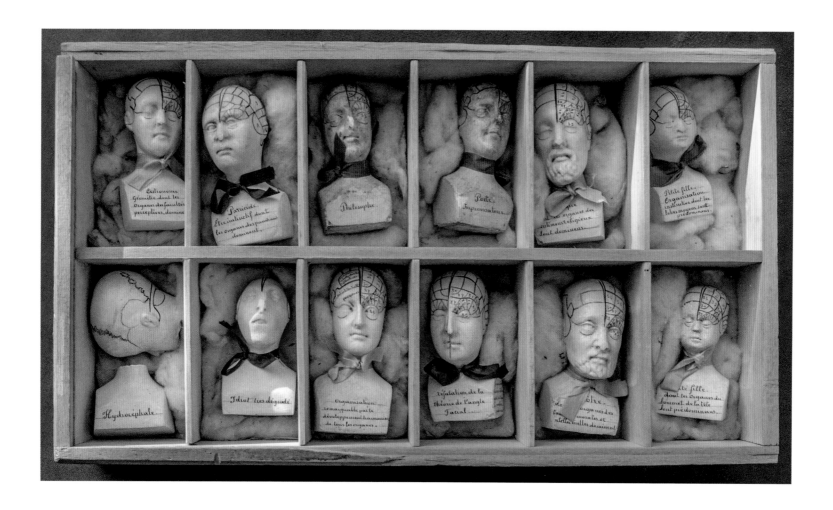

12 phrenology busts
19th century
Each: 8 × 4 × 5 cm
(3⅛ × 1⅝ × 2 in)
Paris

The basic 'laws' of phrenology were laid down by German physician Franz Joseph Gall (1758–1828) in the introduction to his book *The Anatomy and Physiology of the Nervous System in General, and of the Brain in Particular* (1819). Gall studied anatomy in the late 1700s with the aim of deciphering how the shape of the human brain correlated to its function.

Gall set out to link aspects of character to specific locations in the brain and to analyse how damage to certain areas of the brain could result in difficulties in language use, changes in character and other neurological problems. He claimed there were 27 'inner senses' and that the development of these senses determined the shape of the brain, which was then reflected in the shape of one's skull.

Gall devised a method of 'reading' the shape of a person's skull in order to identify the varying development of different cerebral 'organs' responsible for different traits. Phrenologists would run their bare hands over a person's head in order to distinguish areas of elevation, indicating an excess of a certain trait, or indentation, indicating a deficit, sometimes using callipers, measuring tapes and other instruments in the process. They then used these findings to interpret a person's intellectual aptitudes and character traits. Phrenology has been almost universally dismissed as a pseudo-science since the mid-nineteenth century.

This charming box contains 12 curious heads: seven men, two women and three youths. Each has a coloured ribbon around its neck, the significance of which is not clear although it certainly looks like some sort of categorising principle. Their skulls have markings indicating various aptitudes or characteristics such as religious, poetic, philosophical and murderous.

All the faces are quite different: the skull of the mystic (top, second from right) shows dominant religious areas, the parricide (top, second from left) looks surly and has a protruding brow, and the head of the degraded idiot (bottom, second from left) features a backward sloping forehead and receding chin.

It is virtually impossible to imagine how these little heads were used because there probably were so many other variations that they could have devised. The box, which was made in Paris and features a manufacturer's stamp on its bottom, would have enabled this set of heads to be carried around easily.

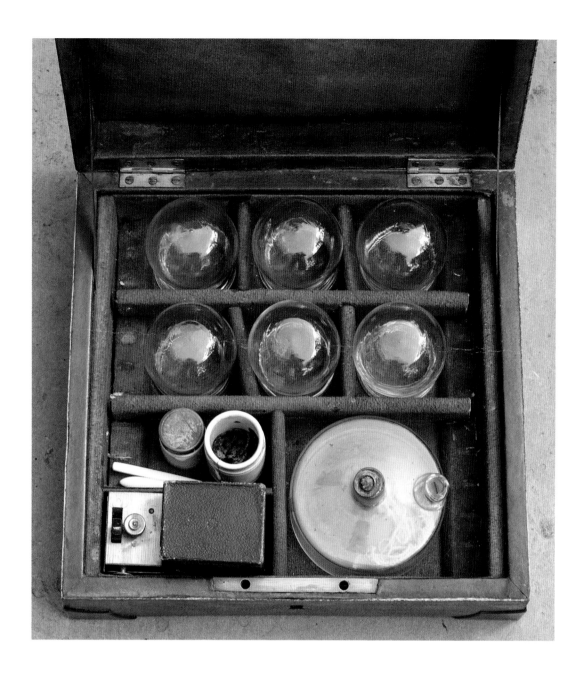

Cupping set
Mid-19th century
27 × 28 × 11 cm
(10 ⅝ × 11 ⅛ × 4 ⅜ in)
France

The medical practice of cupping uses a flame to produce a vacuum in a cupping vessel, which is then applied to the skin. 'Dry' cupping uses only the heated cup, while 'wet' cupping is a form of bloodletting, in which the skin is scarified and blood drawn from the wound by suction. Cupping is documented in early Chinese, Egyptian, Islamic, Native American and Greek cultures, and is mentioned in the work of Hippocrates as a method of drawing excessive humours from the body.

As well as the actual cups and a small burner, this set also includes a small instrument for scratching the skin to allow the drawing out of blood. The box has an outer covering of very fine shagreen, made of either shark or ray skin, a material that seems somewhat incongruous in contrast to the very utilitarian nature of the instruments inside.

As I lay down I saw on a bed nearly opposite me a small, round-shouldered, sandy-haired man sitting half naked while a doctor and a student performed some strange operation on him. First the doctor produced from his black bag a dozen small glasses like wine glasses, then the student burned a match inside each glass to exhaust the air, then the glass was popped on to the man's back or chest and the vacuum drew up a huge yellow blister. Only after some moments did I realise what they were doing to him. It was something called cupping, a treatment which you can read about in old medical text-books but which till then I had vaguely thought of as one of those things they do to horses.

George Orwell, *How the Poor Die* (1946)

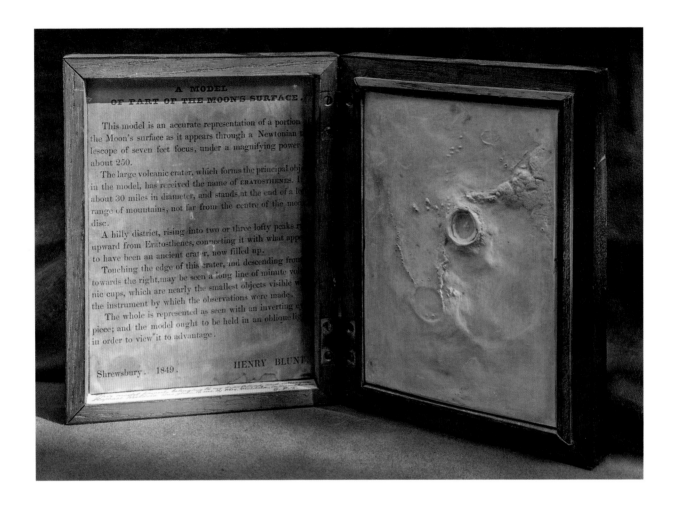

A MODEL
OF PART OF THE MOON'S SURFACE.

This model is an accurate representation of a portion
the Moon's surface as it appears through a Newtonian t
lescope of seven feet focus, under a magnifying power
about 250.

The large volcanic crater, which forms the principal obj
in the model, has received the name of ERATOSTHENES. I
about 30 miles in diameter, and stands at the end of a l
range of mountains, not far from the centre of the moo
disc.

A hilly district, rising into two or three lofty peaks r
upward from Eratosthenes, connecting it with what app
to have been an ancient crater, now filled up.

Touching the edge of this crater, and descending fro
towards the right, may be seen a long line of minute vo
nic cups, which are nearly the smallest objects visible w
the instrument by which the observations were made.

The whole is represented as seen with an inverting e
piece; and the model ought to be held in an oblique lig
in order to view it to advantage.

Shrewsbury. 1849. HENRY BLUN

Henry Blunt
A model of part of the
moon's surface
1849
16 × 12 × 4 cm
(6¼ × 4¾ × 1⅝ in)
Shrewsbury

This charming little diptych shows part of the moon's surface including a large volcanic crater, apparently some 30 miles in diameter, and a lofty range of mountains. It reminds me of the religious diptychs of the Renaissance.

The Whipple Museum in Cambridge University has a virtually identical piece from 1854 that was also made by Henry Blunt (1806–53), however the actual plaster model is mounted the other way up with more or less the same text in the museum's version. The Science Museum in London has yet another version, which was made for the Great Exhibition of 1851.

One wonders what these things were really for. It may be that there was no particular purpose other than to show that it was something that could be done.

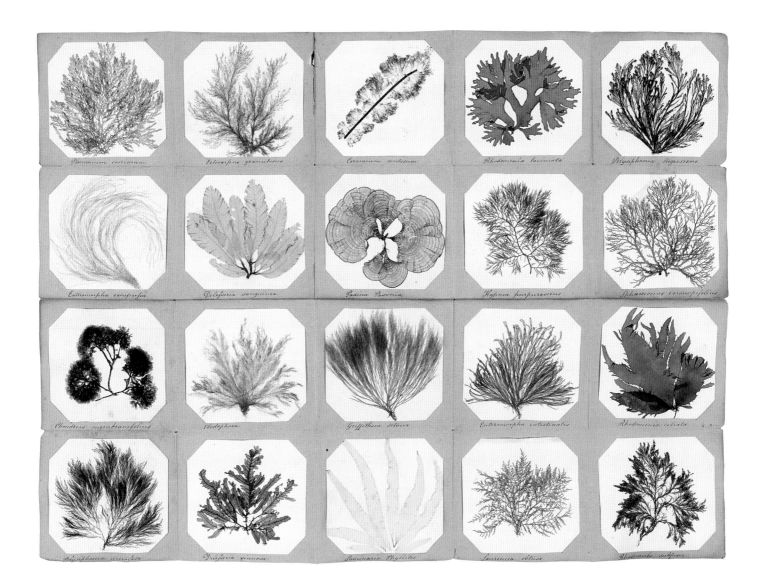

Curtis
Seaweed album
Mid-19th century
15 × 14 × 1 cm
(5⅞ × 5½ × ⅜ in)
Jersey

This small album unfolds to show 20 seaweeds, collected on Jersey in the Channel Islands, with handwritten taxonomic inscriptions and a printed maker's label. It was probably made for interested amateurs visiting the island. Although it looks plain from the outside, it becomes really quite beautiful once opened.

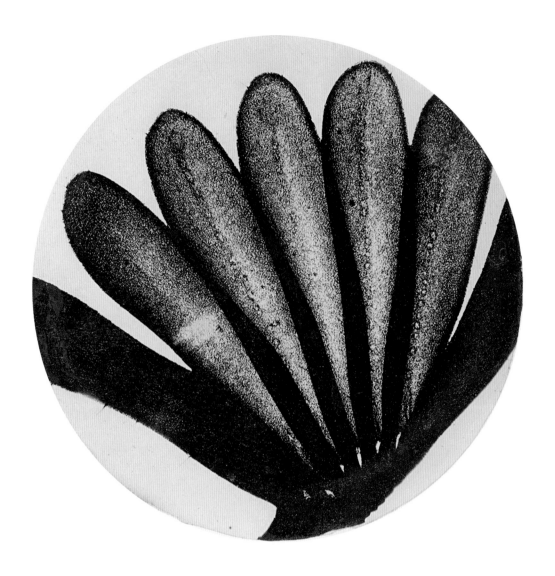

William Henry Olley
The wonders of the microscope,
photographically revealed:
by Olley's patent micro-photo-
graphic reflecting process
1861
36 leaves of mounted
photographs
Height: 25 cm (9⅞ in)
London

While there are various examples of microscope photographs in the collection, this book – produced in 1861 – is the earliest. The introduction to the volume explains that microscope photographs had previously been redrawn by hand onto plates for printing, a process that often led to errors and omissions. Here the photographs themselves are pasted onto the page, an expensive technique used before a process was invented in the 1870s to transfer photographs directly onto lithographic stones or printing plates by chemically sensitising the plates to light.

The photographic images in this book range from parts of animals – such as flies, bees, and spiders –

to plant material and human blood. It is aptly named *The wonders of the microscope* as, rather than conveying any scientific insight regarding its subjects, it offers 'wonder', which can be seen in the image depicting a cross section of the stem of a clematis plant (p.55). The image illustrated here demonstrates the structure of the antennae of a Cockchafer or May bug.

An additional attraction of this book is its *ex libris* as it previously belonged to Solly Zuckerman (1904–93), a well-known scientist who was scientific adviser to the British government and President of the Zoological Society of London.

Preserved conjoined piglets
19th century
30 × 18 cm (11¾ × 7⅛ in)
England

Does such an unusual and visually arresting object require much comment? These piglets are attached forever and hold hands adoringly behind their backs.

This item comes from the Walter Potter sale in 2003. Potter (1835–1918) was a self-taught Victorian taxidermist, who delighted the public with anthropomorphic tableaux of tea-drinking kittens, cigar-smoking squirrels and gambling rats. His museum, originally in Bamber, Sussex, eventually closed in 2003 and the contents dispersed.

At the time that this item was created, the public belief that witchcraft was responsible for this kind of deformation still held some currency. However, scientists had long been interested in mutants – or monsters – as they believed that they might reveal the workings of living beings, and it was this interest that led to the many collections of such mutants that were formed in the eighteenth and nineteenth centuries.

The piglets here, of course, are conjoined twins, meaning they were formed from two individual embryos. This is in contrast to the two-faced kitten in the collection (see p.147), which is likely the outcome of a slight error in the development process of a single embryo, leading to the head widening to such an extent that it ultimately has two faces and almost two heads.

I chose to include an image of the piglets in the first part of the book (p.39) because, looked at objectively, this item suggests a host of concepts: historic, scientific and aesthetic.

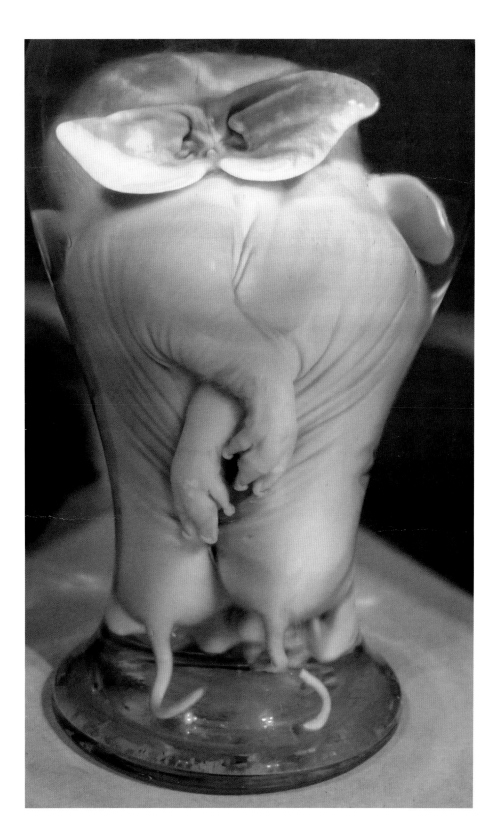

Mineral spar tower
Late 19th century
25 × 7 × 5 cm (9⅞ × 2¾ × 2 in)
Russia

This small tower of mineral crystals was made in Russia, where mining has always been very important. There are two more spar towers in the collection (p.230), however, unlike the others, this one includes the rare feature of an original list of the minerals used.

Spar towers had no didactic purpose, they were intended only to be decorative or as collectors' items for those with an amateur interest in geology and were usually built by miners. In the UK, miners would supplement their income by selling towers like these at annual shows.

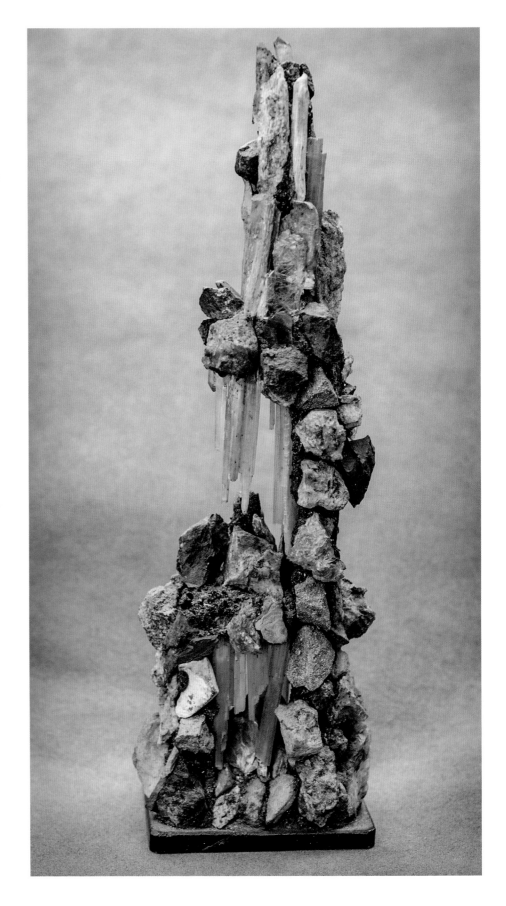

Bisected human skull
19th century
25 × 20 × 8 cm
(9⅞ × 7⅞ × 3⅛ in)
Europe

This skull has been cut in half and veins have been added in wax. The whole is more an object that lies flat than a standing teaching model. Medical students often made preparations like these, apparently for the sheer fun of it.

It would have taken a great amount of force to cut through the bone, but also great delicacy not to damage the specimen and this bisection was clearly done with great skill.

Unlike most anatomical skulls, this one has not been bleached and thus remains much more realistic. The object is both attractive and repellent: it invites tactile appreciation.

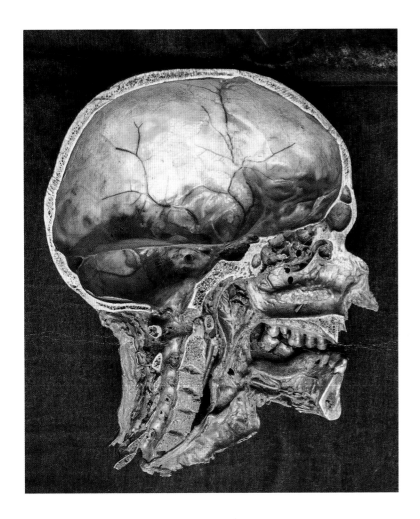

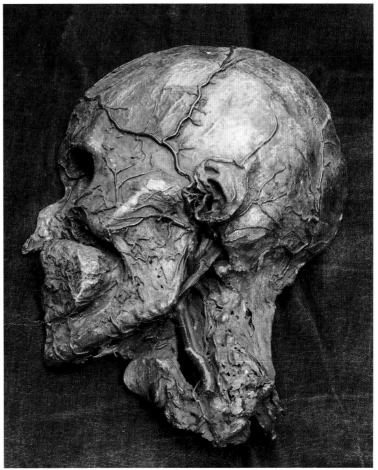

Auguste-Adolphe Bertsch
Photograph of the vertical
cross section of cedar wood
c.1855
34 × 25 cm (13⅜ × 9⅞ in)
France

Auguste-Adolphe Bertsch (1813–71) was another well-known French photomicrographer. His work includes photographs of crystals and insects, as well as wood samples. One of his achievements was to invent the mechanical shutter, which by allowing more control over the exposure of the photographic plate was a marked improvement over the previously used manual flap or pivotal cover.

The Museum of Modern Art in New York has another photograph by Bertsch comparable to this one, of the wood of a chestnut tree, in its collection.

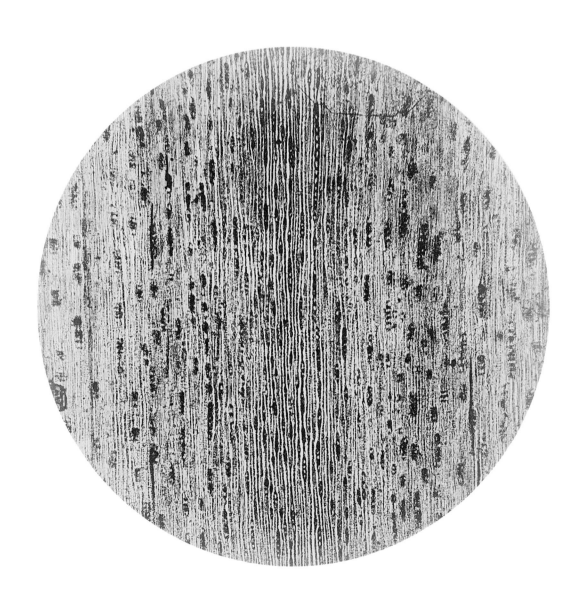

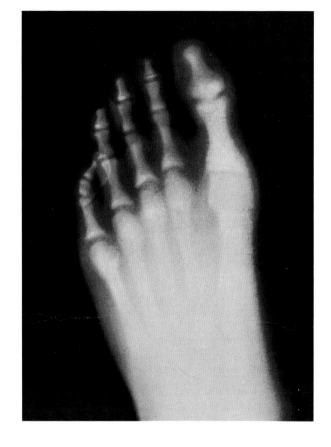

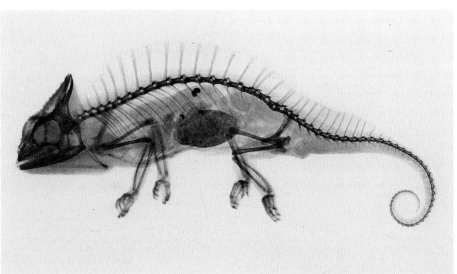

Josef Maria Eder and
Eduard Valenta
Photogravure from X-ray
of *Fuss eines 17 jährigen
Jünglings* and photogravure
from X-ray of *Chamäleon
Cristatus*
1896
Above: 16 × 23 cm
(6¼ × 9⅛ in)
Top right: 28 × 20 cm
(11⅛ × 7⅞ in)
Austria

These are very early X-ray images, one of a chameleon and the other of a foot belonging to a 17 year-old boy. In 1895, while working in Würzburg in Germany, the physicist Wilhelm Conrad Röntgen (1845–1923) discovered X-rays, almost accidentally. This groundbreaking process made it possible to see the interior anatomy of living creatures without the need for surgery.

Josef Maria Eder (1855–1944) and Eduard Valenta (1857–1937) were two Austrian professors in Vienna researching and teaching photochemistry, photometry (the science of measuring light) and scientific photography. Within a year of Röntgen's invention they published a collection of highly detailed photogravure X-ray images, of which these are two examples.

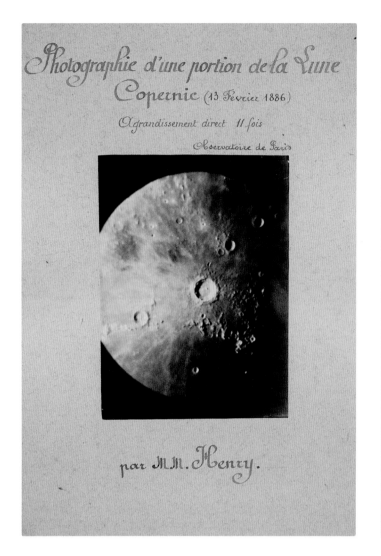

Photographie d'une portion de la Lune
Copernic (13 Février 1886)
Agrandissement direct 11 fois
Observatoire de Paris

par MM. Henry.

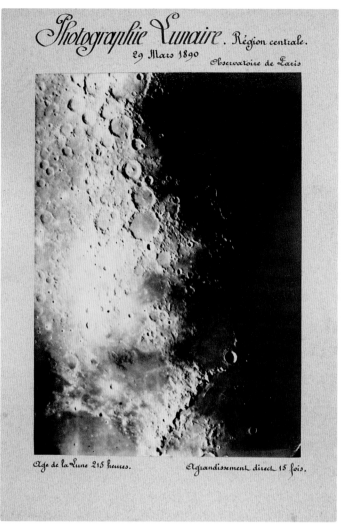

Photographie Lunaire. Région centrale.
29 Mars 1890
Observatoire de Paris

Age de la Lune 215 heures.   Agrandissement direct 15 fois.

Paul and Prosper Henry
Photographie Lunaire: Region Centrale and *Photographie d'une portion de la Lune: Copernic*
1886
Two mounted photographs
Left: 25 × 17 cm
(9⅞ × 6¾ in)
Right: 33 × 25 cm
(13 × 9⅞ in)
Paris Observatory, Paris

These are two photographs taken of different parts of the lunar surface from the Paris Observatory. The Henry brothers who took these photographs – Paul (1848–1905) and Prosper (1849–1903) – were telescope makers and astronomers.

The brothers' photography corresponds to another item in the collection: J Janssen's book (1896; p.191), which tells the story of the observatory and the creation of a new facility at Meudon, outside Paris. Lunar photographs are always intriguing but I found these to be very beautiful as well.

Three glass jars with salt
crystal specimens
c.1852
Each, between:
19–37 × 9–15 cm
(7½–14⅝ × 3½–5⅞ in)
Germany

This collection of cubes in handsome jars, of which
three are illustrated here, has quite a contemporary
feel about it. The salt specimens are probably
from the German town of Strassfurt, where in the
nineteenth century potash – or potassium salts –
was discovered. These became important for
agriculture and manufacturing. Rock or crystal salt
became a by-product for collectors.

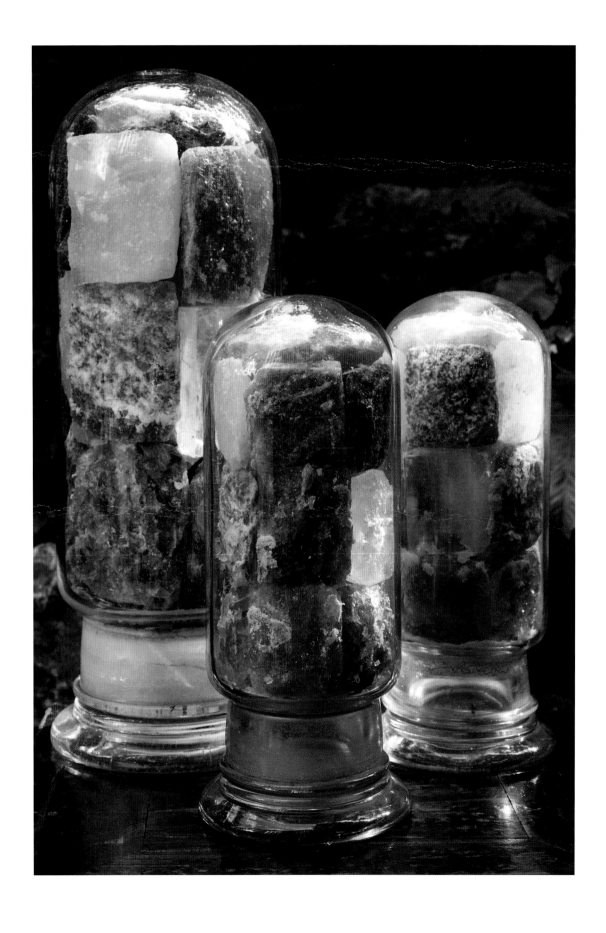

William Booth Grey
Gypsophilia Prostrata (Trailing
Gypsophilia)
Early 19th century
Collage
27 × 20 cm (10⅝ × 7⅞ in)
England

William Booth Grey (1773–1852) was the grandson of the second Duchess of Portland, who was a great collector and very active in the intellectual circles of the late eighteenth century. One of the Duchess's friends was Mary Delany (1700–88), famous for originating flower collages or 'paper mosaics'. These constructions consisted of cutting minute pieces of coloured paper and sticking them on a black background to represent each part of a specimen. Occasionally the pictures were touched up with watercolour paint. Delany's entire oeuvre is now housed in the British Museum in London.

Grey used a similar technique in works such as this. Little is known about him, except that he lived a quiet life, married twice and had no children. Examples of his work are held in the collections of the Metropolitan Museum in New York, the Yale Center for British Art in New Haven and the British Museum.

Grey and Delany's technique remains unique as a way of presenting plant specimens. The result is always aesthetically pleasing, with enough science to prevent it being just decoration.

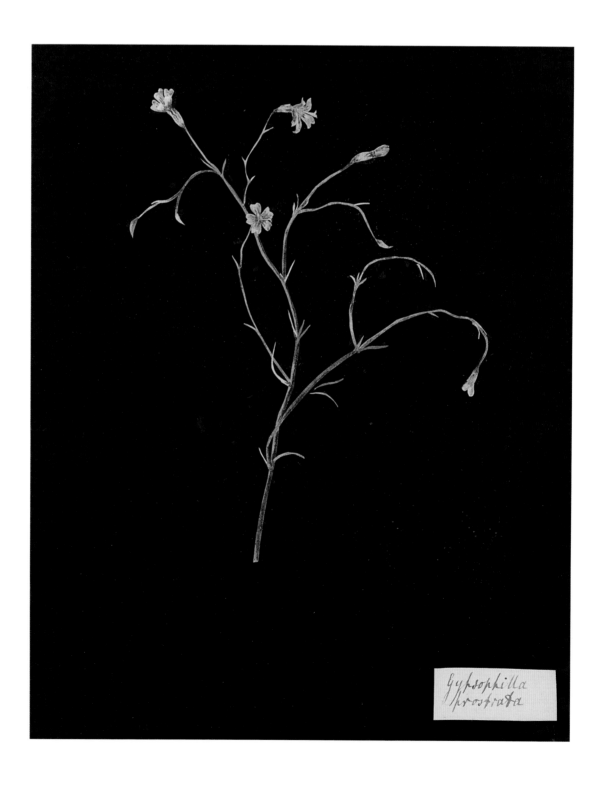

Thomas Hawkins
The book of the great sea-dragons, Ichthyosauri and Plesiosauri, gedolim taninim, of Moses: extinct monsters of the ancient earth: with thirty plates, copied from skeletons in the author's collection of fossil organic remains (deposited in the British Museum)
Frontispiece by John Martin and 30 plates by George Scharf, Henry O'Neill and BJ Rossiter
1840
Height: 55 cm (21⅝ in)
London

Thomas Hawkins (1810–89) was a dealer and collector of fossils. His collection included large marine vertebrates such as plesiosaurs and ichthyosaurs from Lyme Regis on the Dorset coast. He sold much of his early collection to the British Museum, which is now housed at the Natural History Museum where some of the large fossils can still be seen. He wrote two books: *Memoirs of Ichtyosaurii and Plesiosaurii* (1835) and this volume, *The book of the great sea-dragons* (1840).

This volume has 30 plates of skeleton fossils from the author's collection and a frontispiece by the famous English Romantic painter and engraver John Martin (1789–1879). The illustrations are by Henry O'Neill (1798–1880), an Irish artist best known for his lithographs of Celtic art and Irish stone crosses; BJ Rossiter about whom nothing is known; and George Scharf (1788–1860), who is one of my favourite illustrators. Scharf was born in Germany, became an artist, but ended up in the unsettled times during the Napoleonic wars at the battle of

Waterloo in the British army. As a result he moved to England where his knowledge of lithography was unique and worked for scientists including William Buckland, Richard Owen and later Charles Darwin. One of his fine plates is illustrated here.

The Hebrew phrase in the title 'taninim gedolim' refers to a phrase in Genesis 1:21 sometimes interpreted as meaning 'giant reptile', 'giant snake' or 'whales, sea monsters, sea dragons'.

The other well-known, although unrelated Hawkins, Benjamin Waterhouse (1807–94), came from an artistic family and at an early age contributed illustrations to Darwin's *The Zoology of the Voyage of HMS Beagle* (1838–43). Hawkins is most famous for making 33 life-size models of dinosaurs and other extinct animals for the park of the Crystal Palace, rebuilt in south London in 1854, following the Great Exhibition of 1851. The Crystal Palace burnt down in 1936 but Hawkins's large models can still be seen in the park.

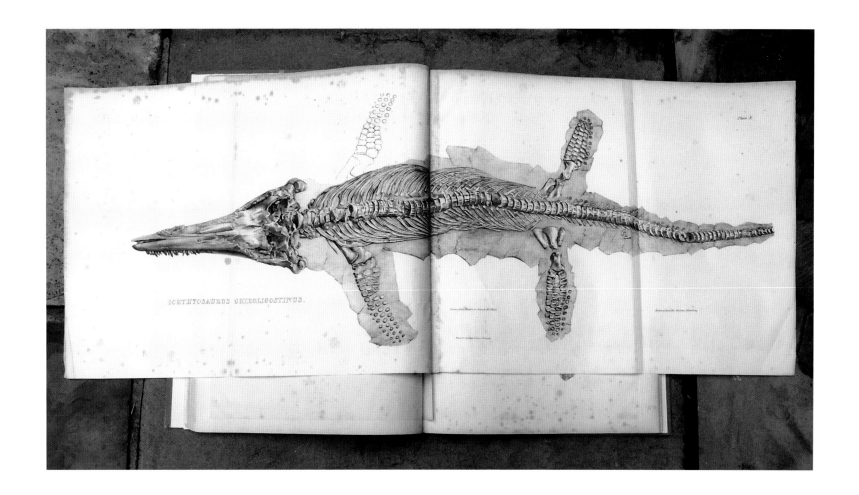

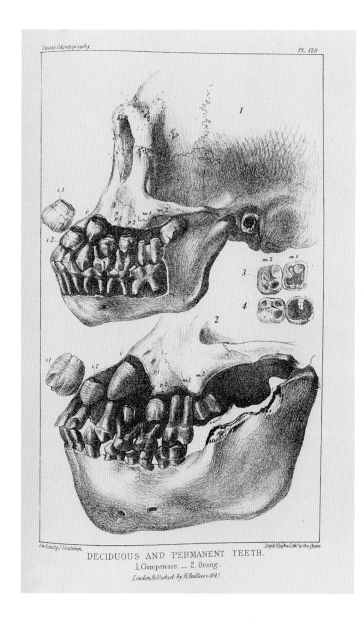

DECIDUOUS AND PERMANENT TEETH.
1, Chimpansee. — 2. Orang.

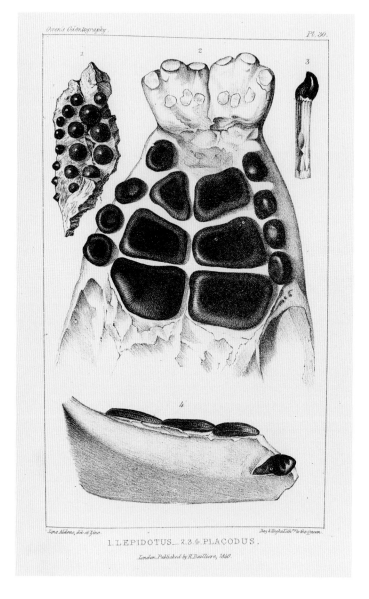

1. LEPIDOTUS. 2.3.4. PLACODUS.

Richard Owen
Odontography: or, A treatise on the comparative anatomy of the teeth: their physiological relations, mode of development, and microscopic structure, in the vertebrate animals
1840–45
168 leaves of lithographs by
J Erxleben, W Lens Aldous
and J Dinkel
2 vols.
26 cm (10¼ in)
London

This publication, Richard Owen's great work on comparative anatomy, is a very technical book, which deals with the development of the teeth of simple and more complex vertebrate animal forms through the ages. Owen (see also p.123) was at the time Hunterian Professor at the Royal College of Surgeons in London.

The second volume is an atlas of 168 plates, drawn by a number of artists, including Joseph Dinkel (1806–91), one of my favourite illustrators. His most famous drawings are arguably those of fossilised fish for Agassiz's *Recherches sur les poissons fossiles* (1833–45; p.235).

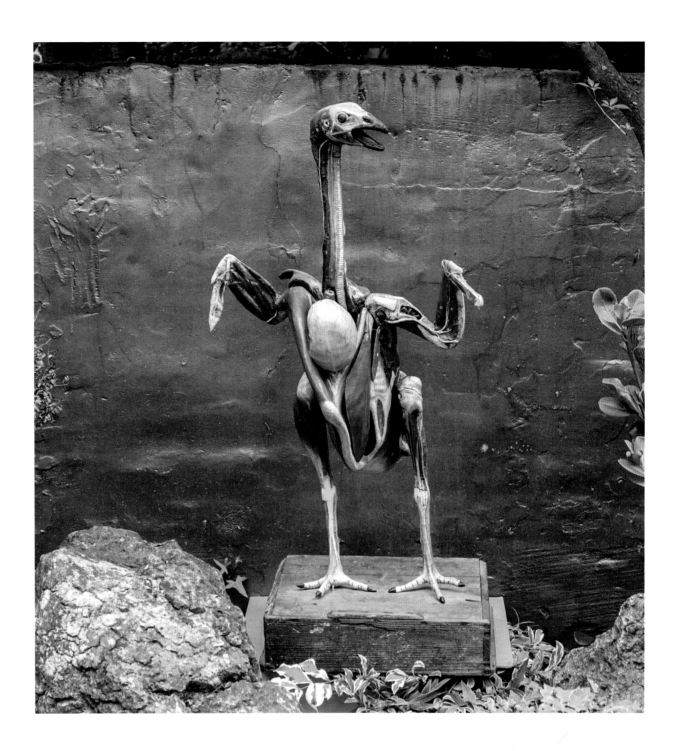

Dr Louis Thomas Jérôme
Auzoux
Papier-mache anatomical
model of a wild turkey
1885
87 × 42 × 37 cm
(34¼ × 16½ × 14⅝ in)
France

This splendid animal comes from an auction of
many Auzoux pieces in Paris in 1998, when an
English dealer, from whom I acquired it in 2004,
bought it.

As with all of Auzoux's anatomical models,
it is accurate and comes apart. The body parts are
numbered and described in an accompanying
booklet. It is rare to find this still with the model.
The Smithsonian Institution in Washington, DC.
has the same model and I was able to give them
a photocopy of the booklet.

This has to be one of Auzoux's most spectacular
models. We haven't had it on the dining room
table yet for Christmas.

Boxed collection of seashells
Early 19th century
11 × 51 × 31 cm
(4 3/8 × 20 1/8 × 12 1/4 in)
England

This collection of shells is housed in two layers and is unlabeled, which suggests the collector was either planning to label them later or was more interested in the beauty of the shells than the science. The collection comes from a period when shell collecting was becoming widespread in Europe as shells from all over the world were brought to England. One of the companies that went on to make up the oil company Royal Dutch/Shell was Shell Trading and Transport, which originally imported shells rather like these to England. It is from these beginnings that the famous scallop shell logo originated.

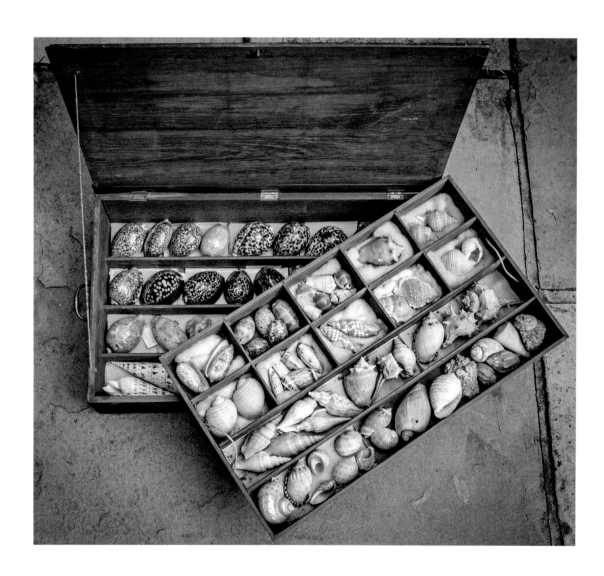

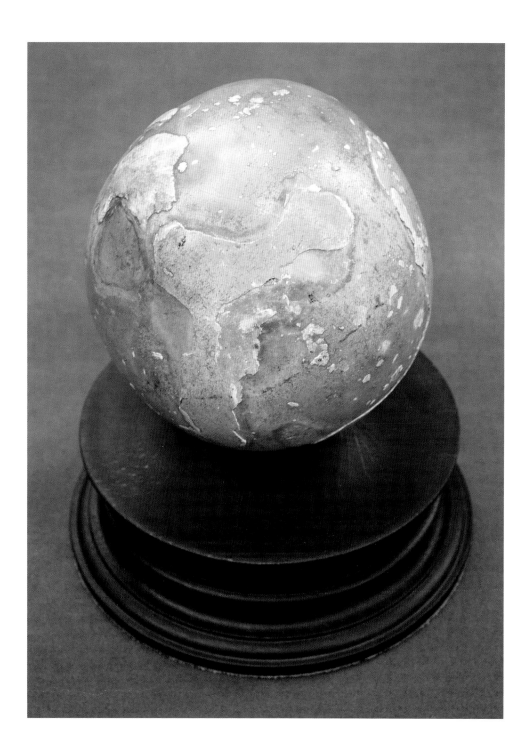

Large bezoar stone
17th or 18th century
21 × 13 cm (8¼ × 5⅛ in)

The word bezoar comes from the Persian *pad-sahr*, meaning antidote, and refers to masses formed from concretions in the stomachs of ruminants. The size of this stone suggests that it came from an elephant or a cow. There was a belief in medieval times that these stones would neutralise any poison and small doses were subsequently taken as a medicine or worn as an amulet.

Bezoars were part of the *Kunstkammer* or Cabinet of Curiosities tradition. Their owners were perfectly happy at the time to believe the myths that were told about them. We now know more and recent examinations have shown that bezoars can indeed remove arsenic-based poisons.

Dr Louis Thomas Jérôme
Auzoux
Four papier-mache models
of mushrooms
Mid-to late 19th century
Each, between:
10–17 × 9 × 9 cm
(4–6¾ × 3½ × 3½ in)
France

These painted papier-mache models were used as simple didactic tools and, unlike Auzoux's more elaborate anatomical models (see pp.143, 151, 157, 178), they do not come apart, which is perhaps not so important in the case of a fungus.

The labels are original and on three of the models read 'comestible', indicating that the mushrooms are edible. The bases are modern. They had probably been handled a lot.

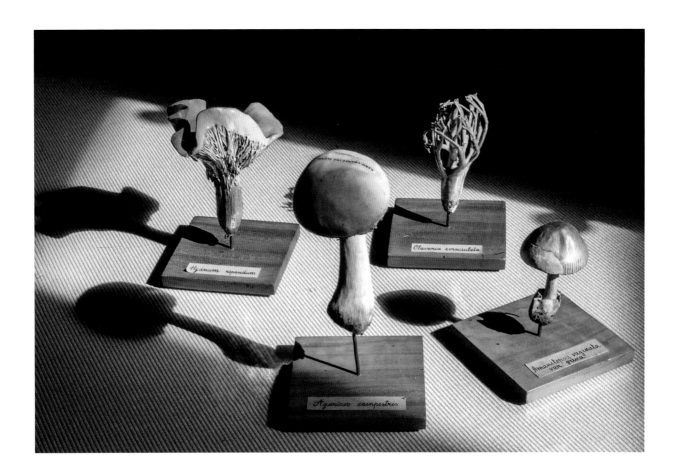

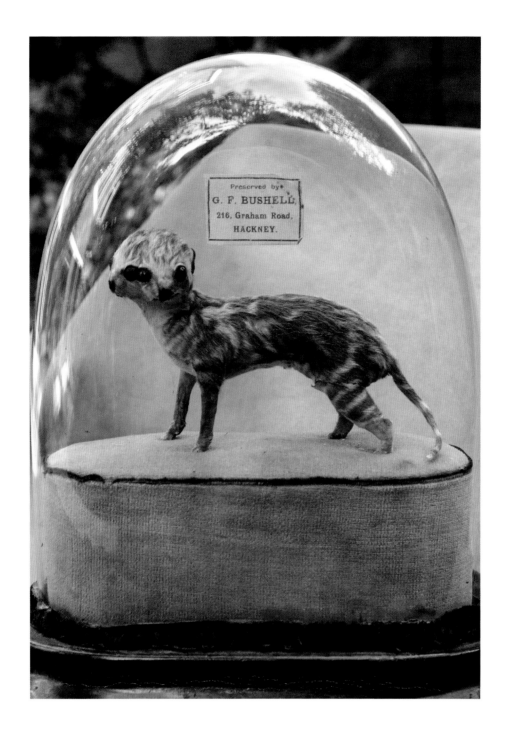

GF Bushell
Taxidermy specimen
of a two-faced kitten
c.1880
25 × 22 × 12 cm
(9⅞ × 8⅝ × 4¾ in)
London

I am particularly fond of this model. One must look
at it very closely to see its attraction, which is why
it is included in the first part of the book (p.47). Its
two heads and four eyes do not become apparent
until you are close to the specimen. The addition of
a delicate velvet cushion is a detail that seems as if
the maker wanted the little beast to be comfortable.
The taxidermist, GF Bushell – of 216 Graham Road,
Hackney, London – stuck his label in a very obvious
position in the glass dome.

I believe this model was made for collections,
rather than for scientific or museum purposes.
The velvet cushion gives it a domestic feel even
though it continues in a tradition of preserving
and studying mutants or monsters.

Three lifecycle specimen jars
Late 19th or early 20th century
Left: 30 × 8 cm
(11¾ × 3⅛ in)
Centre: 45 × 11 cm
(17¾ × 4⅜ in)
Right: 35 × 8 cm
(13¾ × 3⅛ in)
Europe

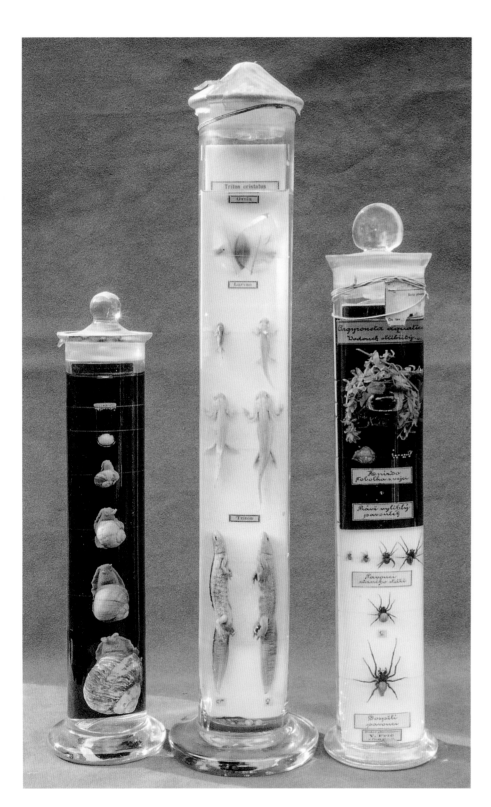

These tall glass jars are typical examples of how specimens were, and still are, preserved for research and teaching use. The largest shows the development of both male and female *Triton cristatus* (crested newts) from the larval stage to maturity.

Contained in the smallest jar is the *Helix pomatia* (escargot), the edible, air-breathing land snail.

The model of the spider is the most interesting of the three. It shows the development of the *Argyroneta aquatica* (diving bell spider), the only spider known to live entirely underwater. It breathes air that it traps in a bubble held by hairs on its abdomen and legs and it builds underwater webs that it fills with air. This process has been very beautifully displayed at the top of the jar.

This last jar comes from Václav Frič in Prague, a Czech naturalist who ran an extensive natural history business throughout the second half of the nineteenth century. Frič came from a prominent family active in law and museum management. He travelled throughout Europe to study how natural history museums were being set up and started his business in 1862. He was soon selling teaching aids and models all around the world and material that he distributed survives in the UK, Austria, Germany, The Netherlands, Italy, Poland and Canada. Frič also acted as retailer for the famous glass model makers Leopold and Rudolf Blaschka (see p.181).

Specimen jar with octopus
Early 20th century
26 × 20 × 10 cm
(10¼ × 7⅞ × 4 in)
Europe

This is in itself not a rare object and this type of specimen can be found in many museum and university teaching collections. However the combination of the specimen and the jar makes for a very aesthetic object. The octopus is suspended in the liquid in such a way as to appear to cling in a lifelike fashion to three nearly invisible wires.

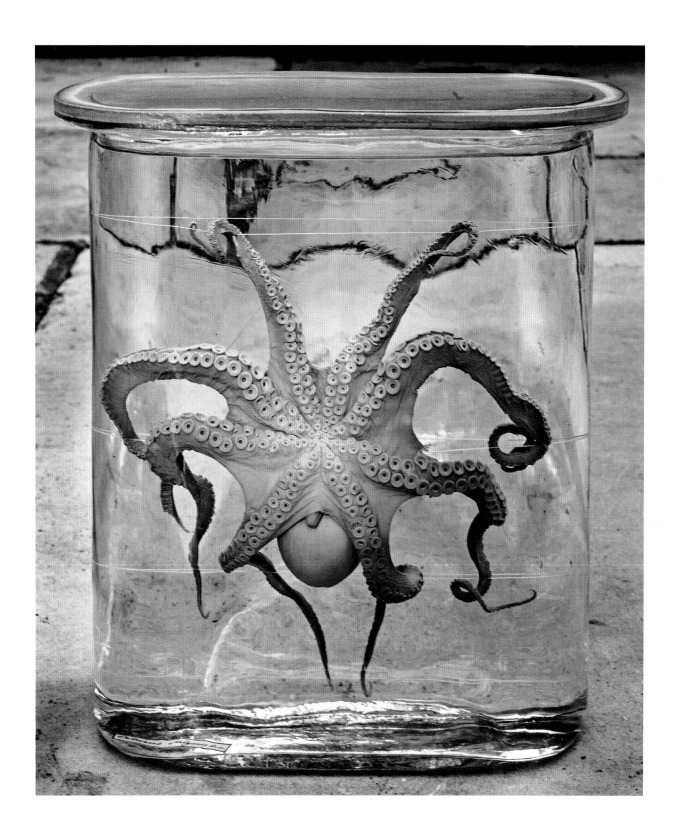

Dr Richard Liebreich
*Atlas der Ophthalmoscopie:
Darstellung des Augengrundes
im gesunden und krankhaften
Zustande: enthaltend 12 Tafeln
mit 57 Figuren in Farbendruck /
nach der Natur gemalt und
erlaeutert*
1863
12 leaves of colour plates
Height: 40 cm (15¾ in)
Berlin and Paris

I find this to be one of the most amazing illustrated books in the collection. It deals with microscopic observations of the fundus, or interior surface, of healthy and diseased human eyes.

Richard Liebreich (1830–1917) was an ophthalmologist who began his career as an assistant to Hermann von Helmholtz (1821–94) in Königsberg, a physician who in 1851 developed the ophthalmoscope. Liebrich's career took him to Berlin, where he worked with the well-known eye doctor Albrecht von Graefe (1828–70) whose memorial still stands (notwithstanding carpet bombing in the Second World War) in the grounds of the Charité Hospital in Berlin.

In 1862 Liebrich went to Paris, where he treated the Empress Eugenie, the wife of Emperor Napoleon III, and then moved to London in 1870 where he continued to practice.

The ophthalmoscope is an instrument that allows one to see into the back of the eye. An initial challenge in its development was the question of how to introduce enough light into this region of the eye. The problem was eventually solved by concentrating a light source through the pupil and observing the back of the eye enlarged through several lenses.

The illustrations were painted by the author directly from his observations and subsequently turned into printing plates, resulting in striking abstract shapes. I find the illustration chosen for the first part of the book (p.51) somewhat reminiscent of late nineteenth-century Japanese kimono design.

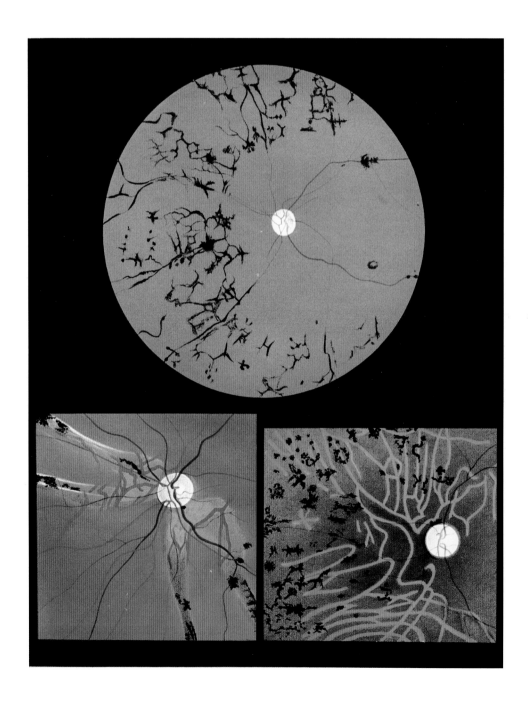

Dr Louis Thomas Jérôme
Auzoux
Papier-mache models of the
male and female
reproductive organs
1885
Left: 43 × 35 cm
(16⅞ × 13¾ in)
Right: 52 × 38 cm
(20½ × 15 in)
France

These models portray the human reproductive organs, and like other anatomical models in the collection, can be taken apart.

The female model, signed by Auzoux and dated 1885, is in very good condition and has clearly not been handled as much as the other model.

The two models came together from a dealer in Brussels. There is of course no certainty that before this they had been together at a teaching establishment, however the contrast in their

condition does raise the interesting question of why this may have come about. Was there less interest in female anatomy? This seems unlikely, because gynaecology was always an important part of the practice of medicine.

The aesthetic appeal of the two models also differs. The female model is not only in better condition but, might one say, it looks rather more... tidy.

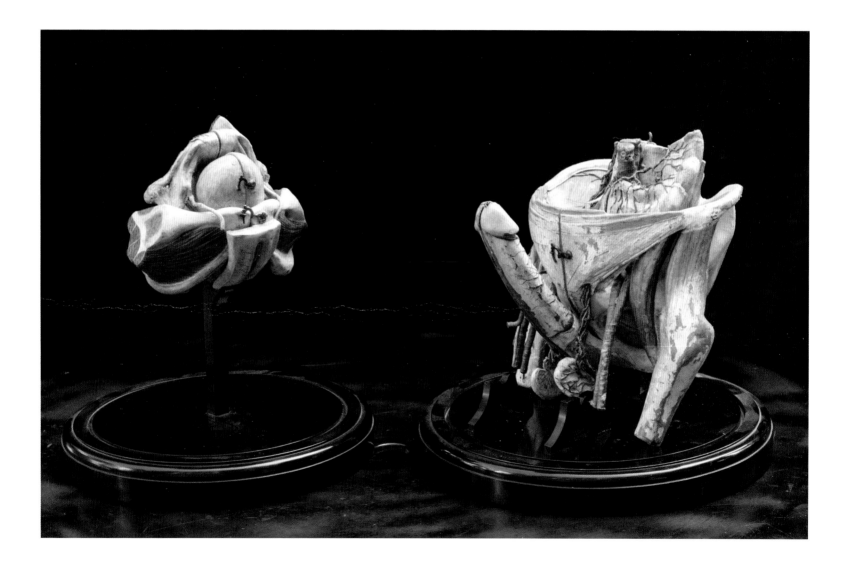

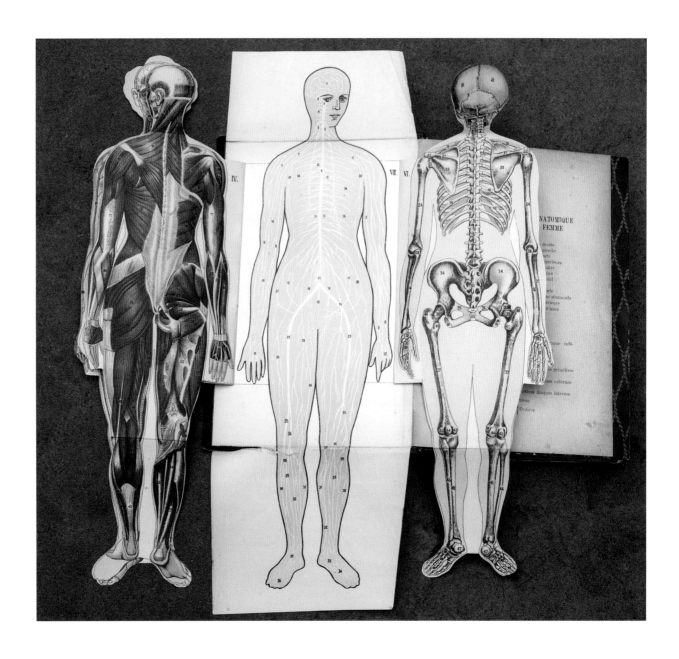

HM Menier
Mon docteur: traite de medecine et d'hygiene, methodes scientifiques et populaires: Allopathie, homeopathie, physiotherapie, medecine pratique, toxicologie, plantes medecinales
1907
4 vols.
Height: 25 cm (9⅞ in)
Paris

This practical guide dates from 1907. It deals with medicine and personal hygiene, describing scientific as well as traditional treatments including alternative medicines, homeopathy, physiotherapy, toxicology and medicinal plants. The plates are fine, if somewhat late in relation to the other illustrated books in the collection.

Volume one has interesting fold-out illustrations of bodies: *Modèles anatomiques démontables*. The text describes various illnesses and treatments, including homeopathic ones, and moves towards more unpleasant illnesses such as herpes, lupus and tumours at the end of the volume.

Volume two deals with diseases of the nervous system, walking problems, male and female sexual diseases, and lastly infantile illnesses.

Volume three deals with lifestyle illnesses such as diabetes, gout and obesity. It then covers infectious diseases, eye, throat and ear problems as well as treatments of various sorts including first aid, advice on certain aspects of cooking, bathing and preparing for military service!

Volume four provides all sorts of practical information about medicinal plants, edible and poisonous fungi, and other poisons. All very French!

Plaster models of teeth
Late 19th century
Left, length: 33 cm (13 in)
Centre, length: 29.5 cm
(17½ in)
Right, length: 27.5 cm
(10¾ in)
Europe

These curious oversized plaster models of teeth were obviously made for teaching as they open up to show their inner structure, a design that makes them appear like very curious sculptural objects. Nothing is known of the manufacturer of these models.

Dentistry organisations and teaching establishments started to appear after the 1850s, marking the start of regulating dentistry. Both the Odontological Society and the College of Dentists, founded in England in 1856 and 1857 respectively, would have needed teaching material such as these models.

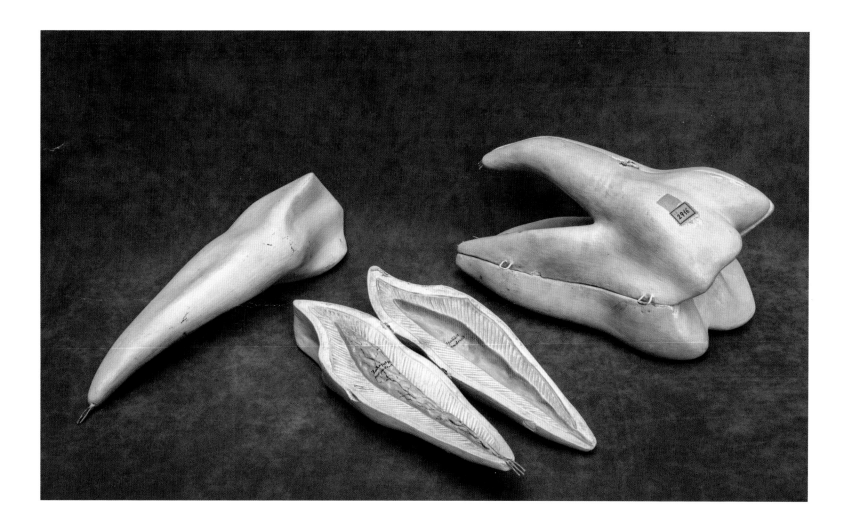

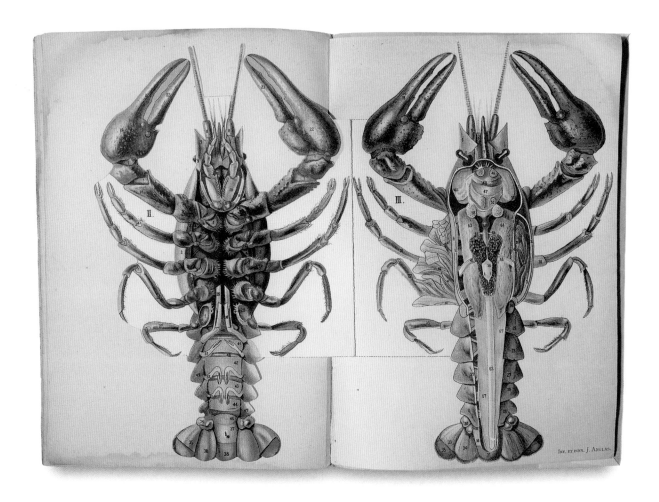

Jules Anglas
*Les animaux de laboratoire:
l'ecrevisse: anatomie et
dissection: trois planches
coloriees a feuillets decoupes
et superposes*
c.1905
Height: 28 cm (11⅛ in)
Paris

I have two books by Jules Anglas (1869–?), about whom very little is known other than his date of birth and the fact that he wrote a significant number of popular illustrated medical and natural history books. He was very keen on illustrating anatomy using 'flap anatomies', showing one layer after another of body parts. This example shows the crayfish using this technique.

Other terms used for these 'flap books' are 'fugitive sheets' or 'moveable books'. The tradition of 'flap anatomies', which began in the late seventeenth century or possibly earlier, mimics in a crude way the act of dissection.

Jules Anglas
*Les animaux domestiques:
cheval, vache, mouton, porc,
chien, coq et oie: exterieur et
anatomie, avec 35 planches
coloriees a feuillets decoupes et
superposes*
1904
7 folded colour prints
Height: 23 cm (9 1/16 in)
Paris

What attracted me to this book about the
physiology and anatomy of domestic animals are
once again the flap illustrations, referred to here as
'*planches ingénieusement disposées*' (plates ingeniously
layered). It is not detailed enough for veterinary
studies, which suggests that the book was aimed
at the general public, but it does contain elaborate
descriptions of the body parts – skeletons, circulation and muscle – as well as organs of animals
including the horse, cow, sheep, pig, dog, chicken
and goose.

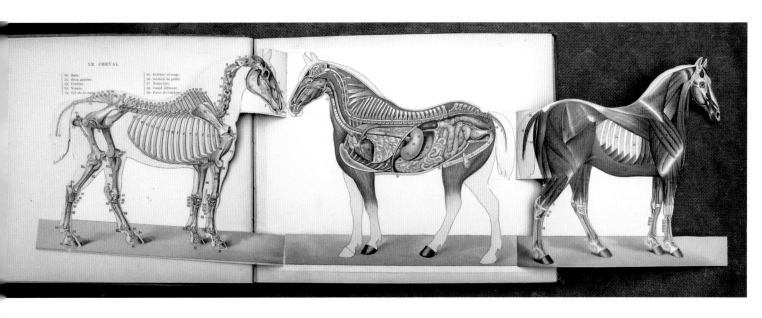

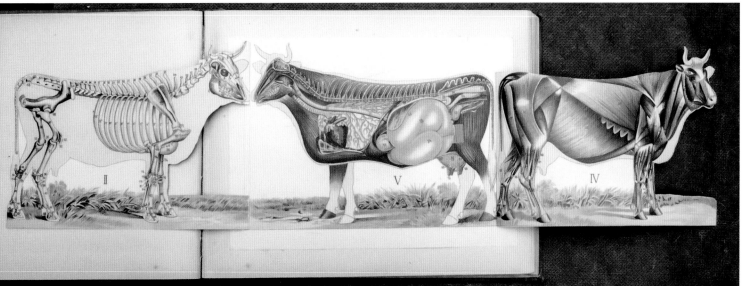

This is the first of a number of very different wax models in the collection, which range from these chick embryos to models of fruits and flowers and pathological and didactic medical models.

Wax was used in both religious and secular art well before the eighteenth century when it began to be used for more scientific and particularly medical purposes. The earliest medical models that I have come across are in the collection of the University of Bologna. There are also very beautiful late eighteenth-century anatomical and botanical models in Florence. Wax continued to be used throughout the nineteenth century and well into the twentieth century, and Madame Tussauds, the museum of waxwork figures, is still a very popular tourist attraction with branches in a number of major cities.

This dish of wax models showing the development of the egg is not a particularly early example, but it has a distinguished provenance (the Royal College of Surgeons of England) and typically reflects the nineteenth-century obsession with embryos: ordered development from unorganised matter. The egg in the centre is definitely not a chicken egg, and is more likely that of a goose.

It is shown here without its glass dome.

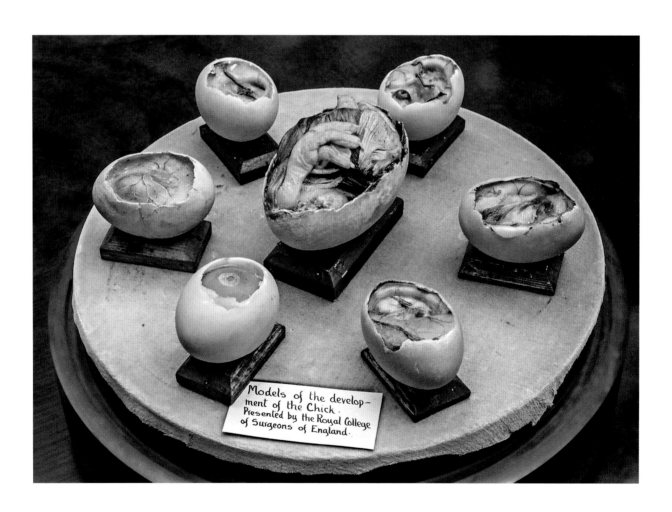

Dr Louis Thomas Jérôme
Auzoux
Papier-mache model
of a snake's head
Late 19th century
28 × 28 × 16 cm
(11⅛ × 11⅛ × 6¼ in)
France

It is not clear what this model is intended to demonstrate because there is no suggestion that it was attached to the rest of the body. However, despite its poor condition, this head successfully conveys the threat of the snake through its open mouth and fierce eyes.

Two other models of snakeheads in the collection (p.187) are made of metal to demonstrate the operations of the upper and lower jaws of the snake, which by disconnecting allow the creature to swallow very large prey.

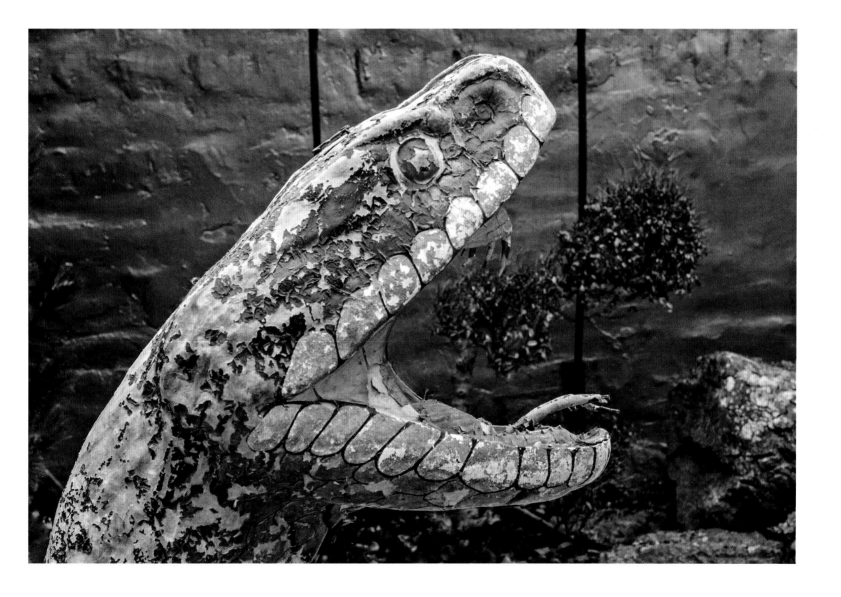

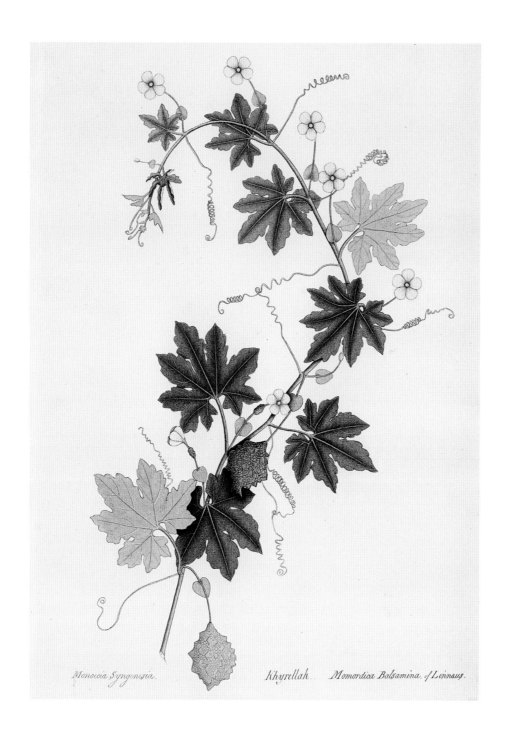

Monoecia Syngenesia.   Khyrellah.   Momordica Balsamina, of Linnaeus.

Two botanical watercolours
Late 18th century or early
19th century
51 × 35 cm (20⅛ × 13¾ in)
India

These watercolours are in the style of the Calcutta School of botanical art, a style initiated by the Scottish surgeon and botanist William Roxburgh (1751–1815) in the early nineteenth century. This example shows a balsam apple.

Indian artists made the watercolours to the specification of their English patrons, using European paper, often from The Netherlands. A considerable amount of this type of work was produced in the late eighteenth and early nineteenth centuries.

Roxburgh was employed by the East India Company and is best known for having run the Calcutta Botanic Garden where he experimented with the adaptation of coffee, pepper and breadfruit to local conditions. By the time he left Calcutta in 1813 he had amassed some 2,500 botanical drawings, most of which can now be found at the Royal Botanic Gardens, Kew, while others are in Calcutta and at the British Museum in London.

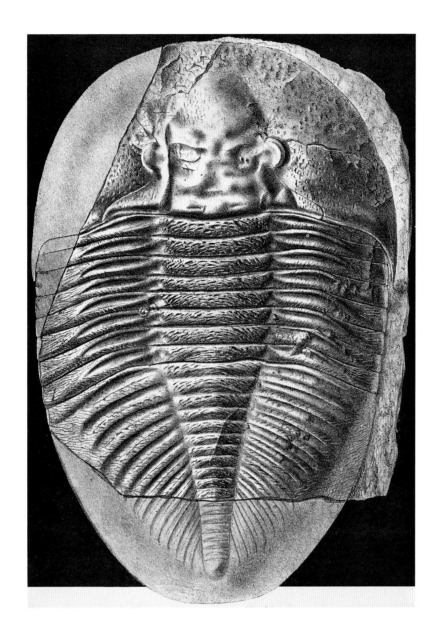

Trilobites are three-lobed marine animals with external skeletons. The name 'trilobite' refers to the three lobes, or segments, running from head to tail that they all share. They were highly successful, widely spread and roamed the seas in many different forms from the early Cambrian period (520 million years ago) to their extinction at the end of the Permian period (250 million years ago). The horseshoe crab is probably their only living relative. Trilobites ranged in size from one millimetre to 78 centimetres, and had eyes, not just single ones but sometimes a complicated combination of lenses.

Until the genus was properly described in the early nineteenth century trilobites were thought to be 'flat fish' or 'petrified insects'.

John William Salter (1820–69) was a naturalist, geologist and palaeontologist who began his career as an illustrator for the mineralogist James de Carle Sowerby (1787–1871), who was part of an extended family of illustrators and whose daughter Salter went on to marry. Salter later worked for Adam Sedgwick (1785–1873), Professor of Geology at Cambridge University. Unfortunately, he committed suicide before he could finish this work on trilobites, so the book only covers half of the species.

The magnificent etchings are by Salter himself together with an illustrator named A Gawan about whom not much is known.

15 paintings of fish
c.1834–41
Each: 32 × 43 cm
(12⅝ × 16⅞ in)
Japan

These fish paintings were made between 1834 and 1841 by two or three artists. The texts are not particularly scientific but do contain instructions on how to cook the fish. Some explain the origins of the names shown in Chinese characters. None of this helps to explain what these illustrations were for.

Fifteen different species are shown and include tuna, bream, sardine as well as a number of species that are too difficult to identify.

Compared to the other Japanese works in the collection, these are relatively early and beautifully produced, some with a metallic shine that is hard to see in reproduction.

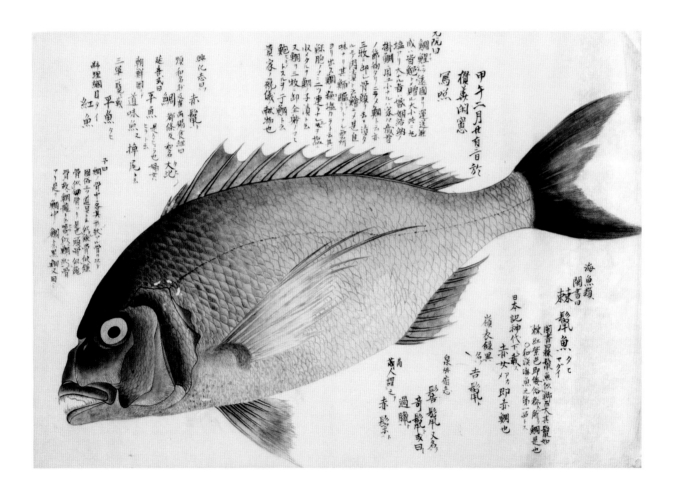

Jacob Davis Babcock Stillman
*The horse in motion, as shown by
instantaneous photography:
with a study on animal mechanics
founded on anatomy and the
revelations of the camera, in which
is demonstrated the theory of
quadrupedal locomotion*
1882
107 plates and 12 line-
engravings, after original
photographs by Eadweard
Muybridge
Height: 32 cm (12⅝ in)
Boston

Jacob David Babcock Stillman (1819–88) and
Eadweard Muybridge (1830–1904) were both
colourful characters. Stillman practiced medicine in
New York, joined the Gold Rush to California in 1849,
then returned to New York and travelled in Europe,
whilst writing copiously about his experiences. He
ended up in California and wrote this book about
Governor Leland Stanford's famous bet on whether
or not a horse's four feet left the ground as it
galloped. The artist for this book was Eadweard
Muybridge, an English photographer who moved
to the United States and became famous there
for his work on animal locomotion, using multiple
cameras. For this project, Muybridge set up a
battery of 24 rapid action cameras each set within
one twelfth of a second to photograph a moving
horse. The book and its illustrations proved that
horses do indeed lift all their feet in gallop.

In France, the physiologist Étienne-Jules Marey
(1830–1904) was doing similar work and using
photography to study movement, particularly the
flight of birds. In 1890 he published *Le Vol des Oiseaux*.
Marey began his career in cardiology and developed
sophisticated physical instruments for measurement, before becoming a pioneer of cinema.

In 1874 Muybridge killed the lover of his wife,
Flora Shallcross Stone, but was acquitted on
the grounds of 'justifiable homicide'. He moved
back to England in 1894 though Flora had died in
the meantime.

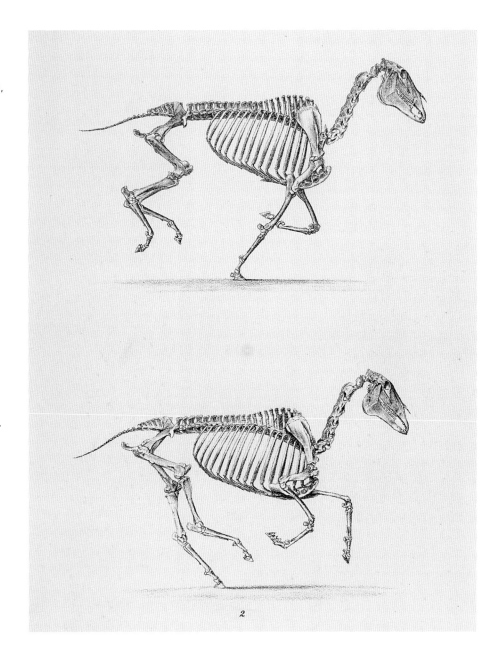

2

Anna Atkins
Cyanotype negative of
Asplenium bulbiferum
(Mother spleenwort)
c.1853
36 × 24 cm (14⅛ × 9½ in)
England

Anna Atkins's (1799–1871) father was a scientist active in zoology and mineralogy at the British Museum, London. Through him in 1839 she met William Henry Fox Talbot (1800–77), one of the inventors of photography, as well as Sir John Herschel (1792–1871), an astronomer, inventor, chemist and experimental photographer who developed the cyanotype process, whereby an item is placed on chemically treated paper and exposed to sunlight. When subsequently washed with water, the paper turns a vivid blue – a phenomenon which led to the term 'blueprint'.

Atkins produced *Photographs of British Algae: Cyanotype Impressions* in three volumes between 1843 and 1853, and subsequently produced two albums of cyanotypes on *British and Foreign Ferns* (1853) and *British and Foreign Flowering Plants and Ferns* (1854).

This cyanotype is of a New Zealand fern, the *Asplenium bulbiferum*, which surprisingly pops up as one of the boxed herbarium specimens in the collection (p.224).

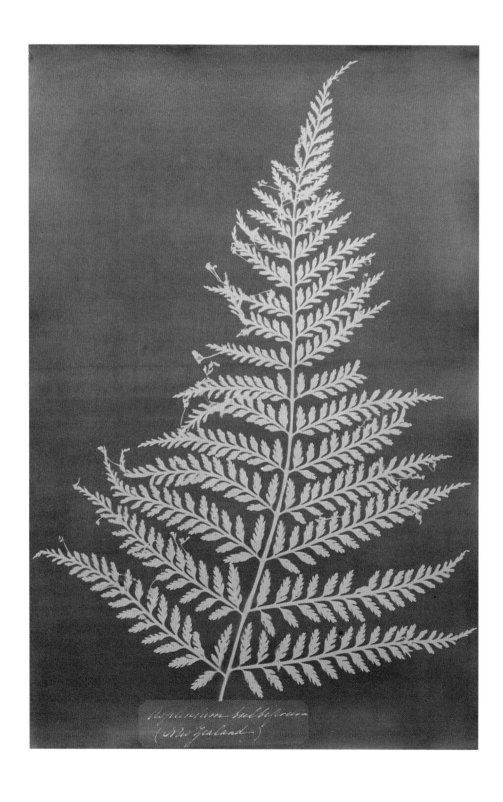

Heinrich Wettstein
Wandtafeln für den Unterricht in der Naturkunde: Botanik
1871
35 plates
83 cm (32⅝ in)
Zurich

This book is a bound set of teaching diagrams, dealing with botany and produced in a large format for use in classrooms. These diagrams, dating from 1871, were printed; whereas earlier teaching diagrams would usually have been hand-painted.

It is still possible to find some of these nineteenth-century teaching aids in storages at numerous universities. Often the professors who used them were quite accomplished artists in their own right. At Cambridge University, for example, Darwin's botany teacher John Stevens Henslow (1796–1861), the man who got him the job on

HMS *Beagle*, made amazing teaching diagrams, some 60 of which are preserved in storage at the Whipple Museum of the History of Science in Cambridge, where they are waiting to be seen.

The purpose of the illustrations was not to demonstrate a particular plant species, but to explain different types of organs. This is why the images have such a modern, abstract feel to them. They are amongst my favourites (see p.12).

I was at a book fair in London when this volume, hidden in a stack against the wall of the dealer's booth, called out to me.

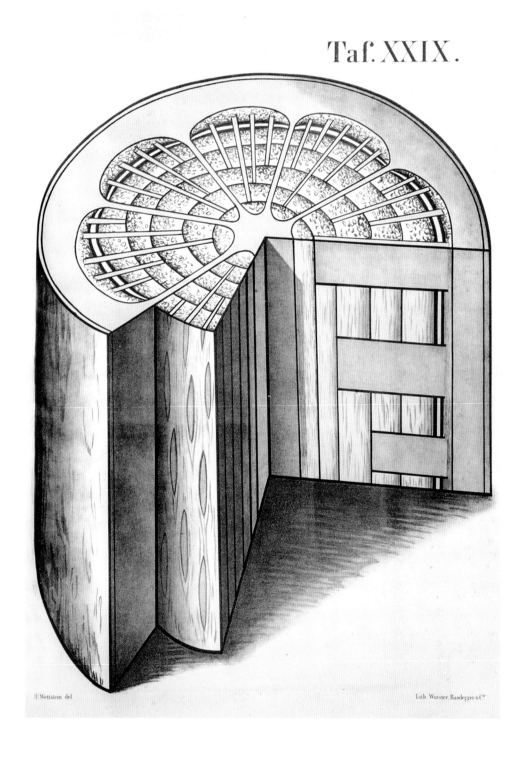

Taf. XXIX.

H. Wettstein del.

Lith. Wurster, Randegger & Cie.

George Young
A geological survey of the Yorkshire coast: describing the strata and fossils occurring between the Humber and the Tees, from the German ocean to the plain of York
1822
22 plates by John Bird
Height: 28 cm (11⅛ in)
Whitby

George Young (1777–1848) was a Scottish pastor, educated at Edinburgh University where he studied mathematics and natural philosophy under John Playfair (1748–1819), amongst others. Playfair was known for his ability to explain the somewhat abstruse ideas of James Hutton (1726–97), a figure who is held by many to be the 'father of modern geology'.

Young became a Presbyterian pastor in Whitby in Yorkshire, where he lived for 42 years and where he founded the Whitby Museum.

Although he was a very good geologist, he was careful about drawing conclusions, particularly if they went against traditional church dogma. In 1838 he published *Scriptural Geology*, in which he attempted to reconcile geology with the teachings of the Bible.

This book on the geology and fossils of the Yorkshire coast is very detailed and really quite early in terms of the history of geology. The illustrations, made by John Bird, who was curator of the Whitby Museum, are noteworthy, showing fine examples of shells and ammonites somewhat naively drawn.

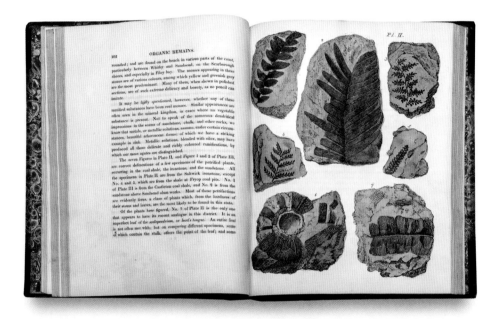

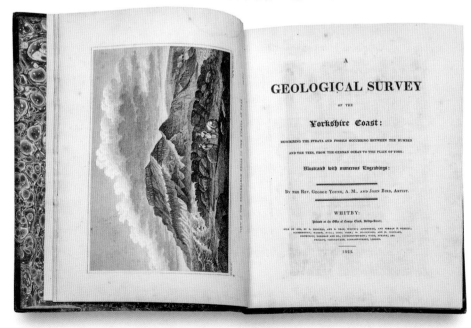

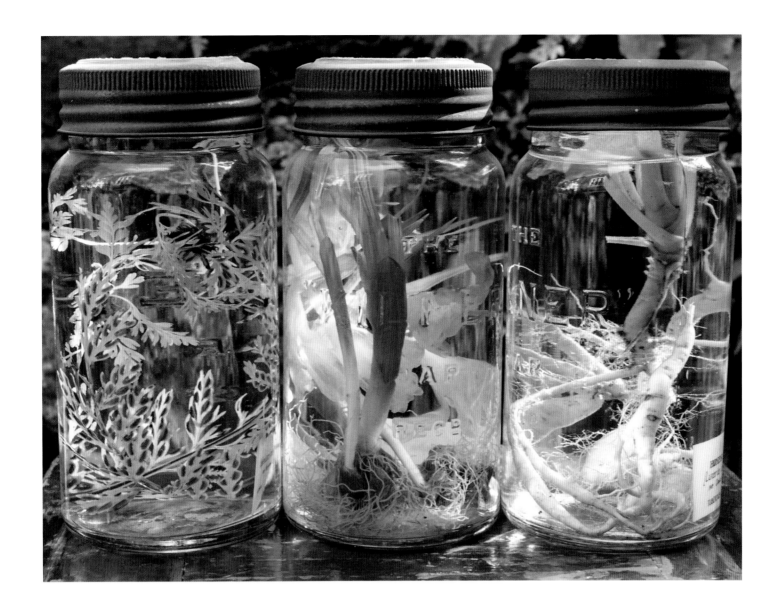

Ten jars of pickled fruit
and vegetables
c.1930
Each: 18 × 10 cm (7⅛ × 4 in)
England

These ten jars are likely to date from the 1930s. In that sense they are later than most of the material that make up this collection. I was drawn to them when I saw them at first but was subsequently put off by their late date. Then I saw the photographs Rosamond Purcell had taken of these and it made me look again.

   With very simple means whoever put these specimens together made something beautiful, demonstrating to me yet again that by looking carefully beauty can be found in unexpected places. It is for this reason that they feature in the first part of the book (pp.24–25).

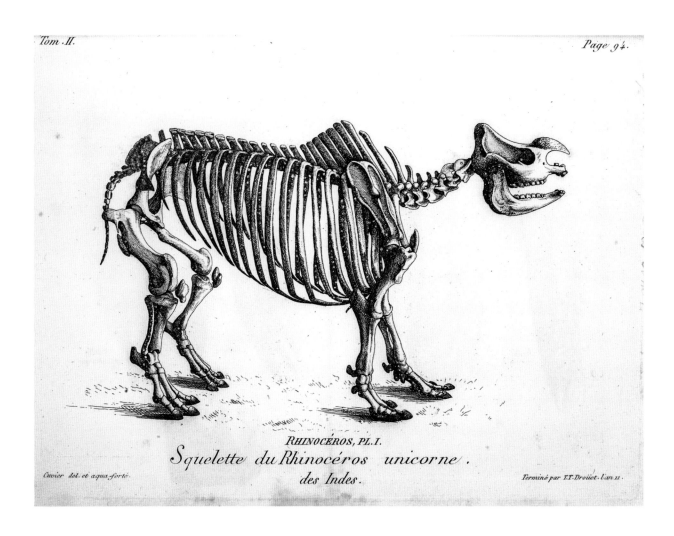

RHINOCÉROS, PL.I.

Squelette du Rhinocéros unicorne.
des Indes.

Cuvier del. et aqua-forte.

Terminé par T.T.Drouet.l'an 11.

M le Baron G Cuvier
*Recherches sur les ossemens
fossiles: où l'on rétablit les
charactères de plusieurs
animaux dont les révolutions du
globe ont détruit les espèces*
c.1821–25
10 vols.
Selected illustrations by
Mlle Balzac, Beuevelot,
Huet, C Laurillard and
N Marechal
Height: 31 cm (12¼ in)
Paris

*Recherches sur les ossemens
fossiles : où l'on rétablit les
charactères de plusieurs
animaux dont les révolutions du
globe ont détruit les espèces*
1825
5 vols.
Height: 31 cm (12¼ in)
Paris

The collection includes two editions of this very important book by Baron Georges Cuvier (1769–1832), dealing with the fossil remains of four-legged creatures and their extinction: the second or 'nouvelle edition' (1821–25) and the third edition (1825).

Cuvier was a naturalist and anatomist at the Muséum National d'Histoire Naturelle in Paris, previously known as the Cabinet du Roi in the Jardin des Plantes. He was instrumental in establishing the fields of comparative anatomy and palaeontology, and was the first scientist to present convincing evidence for the extinction of species. His vast collection of skeletons of all sorts of creatures at the museum in Paris is still well worth visiting.

This is a major work and very comprehensive. It includes fine engraved and lithographic illustrations, many of which are unattributed. A number are by Cuvier himself. Arguably the most interesting illustrations are the work of Huet although it is not totally clear which Huet this was. It seems most likely that it was Nicolas Huet (1770–1830) who worked at the Muséum National d'Histoire Naturelle and also made animal and plant drawings for the

Empress Joséphine (1763–1814), the first wife of Napoleon I.

Huet is notorious today for having made an anatomical study of Sarah 'Saartjie' Baartman (1789–1815), the so-called Hottentot Venus, who was paraded in London and Paris as a 'freak show' attraction. Cuvier actually made a plaster cast of her body before dissecting it. Her story became a focal point for abolitionist protests and a symbol of Western racism. In 2002, at the request of the South African government, her remains were returned to her place of birth and interred.

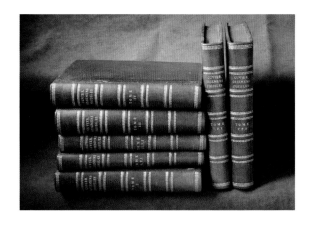

W Breidenstein
Mikroskopische Pflanzenbilder:
sehr starker Vergrosserung zum
Gebrauche bei dem Unterrichte
in der Botanik, nebst einem
Grundriss der Anatomie und
Physiologie der Pflanzen zur
Erlauterung der Abbildungen
1856
16 colour lithographs by
M Frommann
Height: 29 cm (11⅜ in)
Darmstadt

This curious book (roughly translated as 'Microscopic Images of Plants: Very Strongly Enlarged for Use in Botany Lessons') was written for German students of botany who might not have had access to decent microscopes.

The lithographic illustrations of microscopic botanical observations cover material such as cell structure, vessels, bark, sexual organs and pollen.

The illustrations are so stylised that they become incredibly abstract and modern, which of course is what attracted me to the book (see p.43).

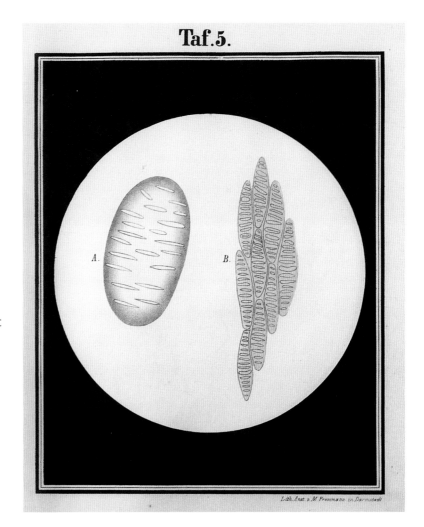

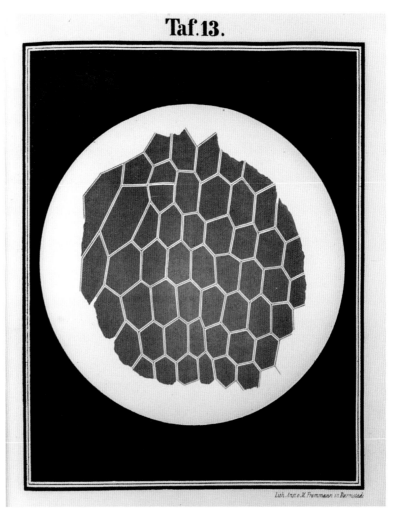

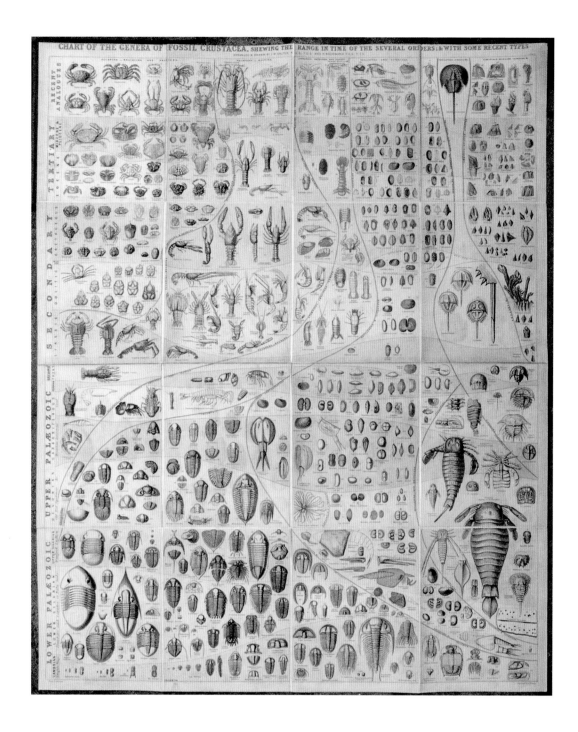

John William Salter and
Henry Woodward
*A descriptive catalogue of all the
genera and species contained in
the accompanying chart of fossil
crustacea, showing the range in
time of the several orders; with
some recent types: illustrated
by upwards of four hundred and
ninety figures*
1865
A single folded leaf of
engravings
Height: 23 cm (9⅛ in)
London

This is a foldout work where a large illustration
unfolds into 16 times the size of the book, showing
fossil crustacea and their distribution over time.
Each fossil is then individually illustrated and
categorised.

The stunning illustrations are by John William
Salter himself, who was also responsible for the
trilobite images (p.159).

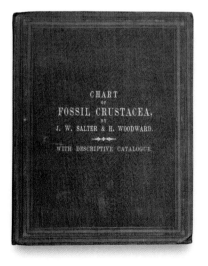

*Transactions of the Linnean
Society of London*
1817–44
17 vols.
Each, height: 31 cm
London

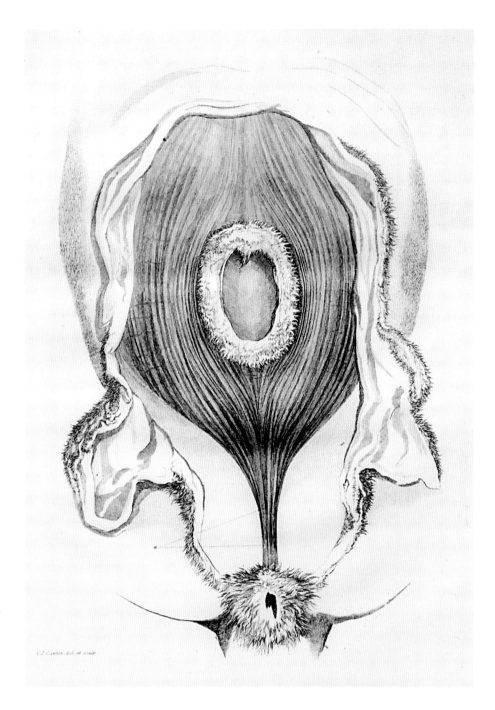

These 17 volumes ranging from 1817 to 1844 are fairly random sets of transactions of the proceedings of the Linnean Society in London from the first half of the nineteenth century. The Linnean Society was founded in 1788 when the collection of plants and fish specimens assembled by the Swedish naturalist Carl Linnaeus (1707–78) was bought and shipped to England. The society is still active to this day.

What attracted me to these volumes initially was the fact that they still had their original boards, which is the format that many publications were delivered to the purchaser in. It was then up to the purchaser to have them bound as desired.

It is ultimately, however, the illustrations which enchanted me. They are by a large number of artists including my favourite George Scharf as well D Allen, CJ Canton, I Curtis, E Farrer, L Guilding, JH Newton, JDC Sowerby, CH Smith, amongst others. John Curtis (1791–1862) made a large number of the illustrations, particularly in the earlier volumes, but individual contributors often used their own favourite artist. The biologist Richard Owen made a habit of commissioning Scharf.

Curtis was an entomologist and illustrator whose most famous work was *A Guide to the Arrangement of British Insects* (1829). His original illustrations are now in the Natural History Museum, London and his insect collections were ultimately sold to the Museum Victoria, Australia and the Natural History Museum, Dublin.

Amongst the competent line illustrations supporting the science being described, one suddenly also finds beautiful hand-coloured illustrations.

My eye was particularly drawn to the illustrations by CJ Canton (?–1840) – an anatomical illustrator employed by Guy's Hospital, London – produced in 1829 as part of a paper on the mammary organs of the kangaroo.

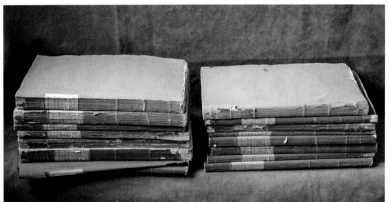

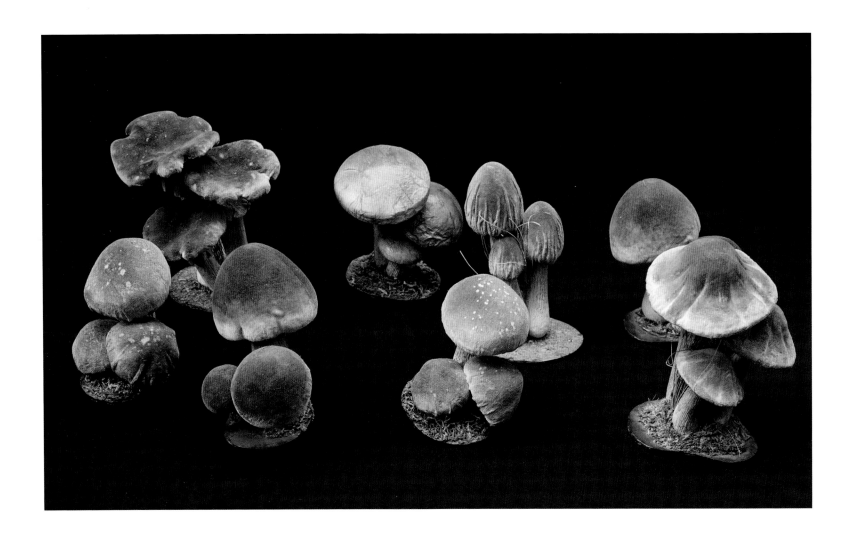

Eight models of mushrooms
Late 19th century
Each, between:
13–20 × 10–15 × 9–12 cm
(5⅛–7⅞ × 4–6 × 3½–4¾ in)
France

Very little is known about these mushrooms. They are very finely made using a number of different materials including velvet. They have no labels, which suggests they were not made for teaching or for showing poisonous varieties. Nothing is known of their provenance. All I can say is that they are very beautiful, possibly more so than the real thing (pp.48–49).

Two preserved soft-shelled
turtles
Late 19th century
Jar: 37 × 28 × 12 cm
(14 ⅝ × 11 ⅛ × 4 ¾ in)
Europe

These two turtles contained in a jar with a preser-
vative liquid are rather enchanting. Their heavy
shells are suspended in the liquid by a very thin line.

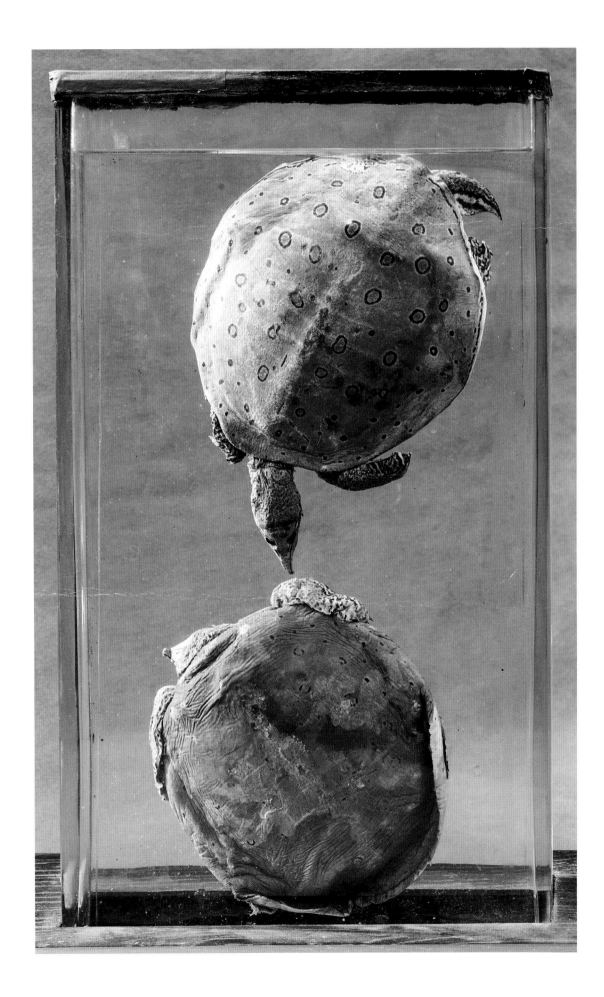

Warren de la Rue
The Bakerian lecture on the total solar eclipse of July eighteenth, 1860: observed at Rivabellosa, near Miranda de Ebro, in Spain
1862
14 leaves of plates:
photography by S Russell,
engraving by J Basire, and
lithography by V Brooks
Height: 32 cm (12⅝ in)
London

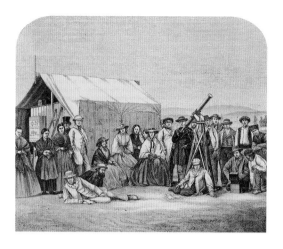

Warren de la Rue (1815–89) was a member of a printing family, who went into the family firm but eventually became more interested in chemical and electrical research. He then took up astronomy and photography. Rather than using daguerreotypes he preferred the wet-collodion process to produce photographic plates. This was the preferred technique until the 1880s when glass plates with emulsion replaced wet plates.

In 1851 a daguerreotype of the eclipse of the sun had been shown to great acclaim at the Great Exhibition in London. De la Rue consequently decided to travel to Spain to photograph the eclipse of 1860.

De la Rue made stunning lunar pictures and turned to photographing the sun in c.1860, using what he called a heliograph that he had devised. This was a sort of telescope created especially for dealing with the highly intensive light. Today a heliograph is a signalling appliance using mirror-directed sunlight reflections, used extensively in the military.

This report to the Royal Society, printed as part of the Society's journal *Philosophical Transactions*, includes a highly technical text. The plates on the other hand are magical. Large lithographs copied from photographs were coloured by hand to demonstrate the red flames seen during the solar eclipse (see p.35). The publication also includes a number of photographs pasted onto paper, an intermediate technique used until methods were discovered to transfer a photograph directly onto a printing plate.

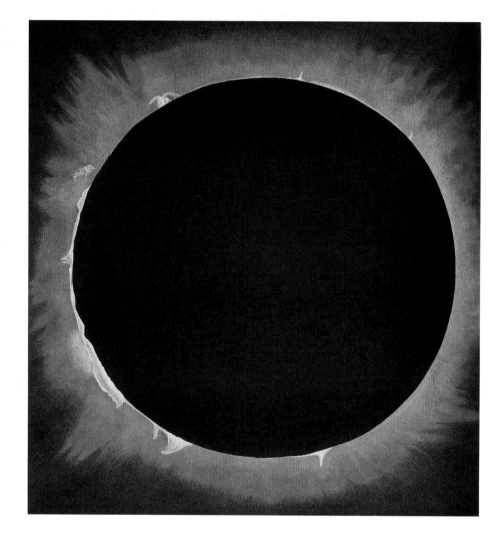

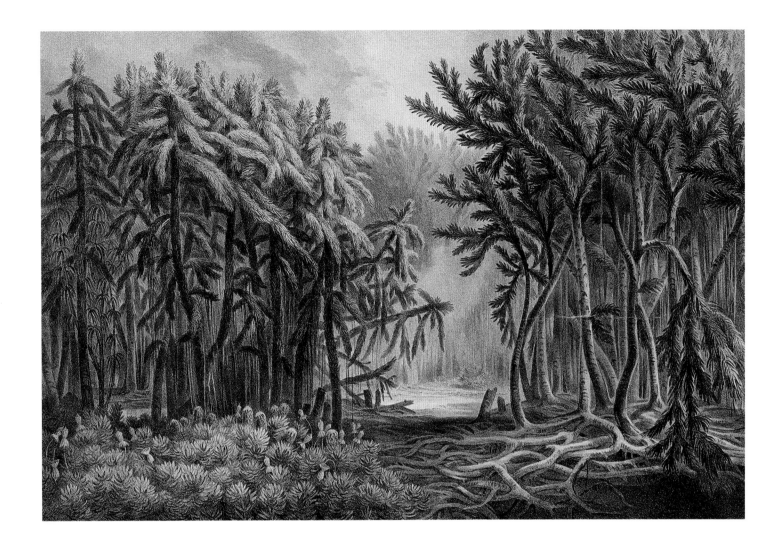

Franz Unger
Die Urwelt in ihren
verschiedenen Bildungsperioden:
14 Landschaftliche Darstel-
lungen mit erläuterndem text
1858
16 loose leaves of
lithographs by Charles
Joseph Kuwasseg
Height: 51 cm (20⅛ in)
Leipzig

Franz Unger (1800–70) was an Austrian doctor and also a professor of plant physiology in Vienna in the 1850s. He developed various ideas on fossils, cell biology, the pathology and distribution of plants, and evolution.

The first edition of this work (translated as *The Primeval World In Various Developmental Periods*) was published in Germany in 1851. To give an idea of what was going on at the same time it is worth noting that Charles Darwin published his famous work *On the Origin of Species* in 1859.

This edition, which features 16 large unbound plates, dates from 1858. Each plate shows a scene from a geological period, right up to relatively modern times, and has an accompanying sheet of explanatory text. The artist was Charles Joseph Kuwasseg (1802–77), another Austrian from a family of artists. Kuwasseg was chosen after

Unger had been put under considerable pressure by his students to try to show visually what he meant when he described various stages of the Earth's evolution. In the introduction Unger admits to the usefulness of visuals in explaining his ideas, but also stresses the limitations of his own knowledge and therefore of the pictorial representations.

The images were a great success and were used for magic lantern slideshows throughout the 1850s.

This is a significant publication because it demonstrates the importance, both then and now, attached to using visual material to convey scientific concepts. At the same time many scientists were reluctant to commit themselves to the depiction of a reality about which they theorised, but of which they could not guarantee the pictorial accuracy.

Edward Hitchcock
Ichnology of New England:
a report on the sandstone of
the Connecticut Valley,
especially its fossil footmarks,
made to the government
of the Commonwealth of
Massachusetts
1858
60 leaves of lithographs
by LH Bradford & Co
Height: 32 cm (12⅝ in)
Boston

Supplement to the Ichnology of
New England: a report to the
government of Massachusetts,
in 1863
1865
20 plates of photograpy
by JL Lovell and lithography
by A Meisel
Height: 32 cm (12⅝ in)
Boston

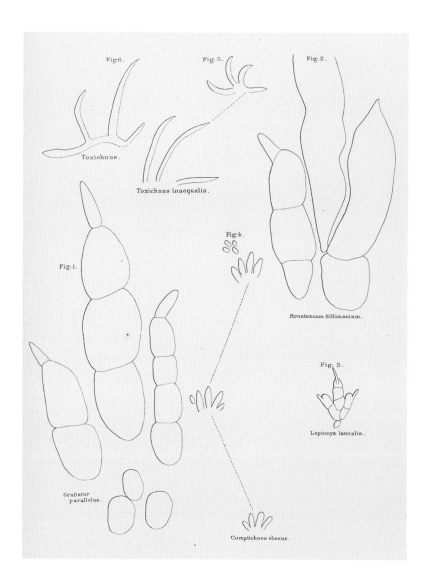

Ichnology is the study of features in the landscape caused by living organisms, such as burrows, trackways and footprints.

Fossilised footprints were discovered in 1802 in the Connecticut Valley in Massachusetts. Initially they were attributed to birds and became referred to as 'Noah's Ravens'.

Edward Hitchcock (1793–1864) acquired the fossilised footprints in the 1830s. Hitchcock was originally a pastor but became a Professor of Chemistry and Natural History, and subsequently of Natural Theology and Geology, at Amherst College. In 1841 he published a first set of identifications of the fossil footprints. He expanded this in 1848 and in 1858 he felt the need to revise the entire work and publish *Ichnology*, a work in which he attributed some of the footprints to ancient reptiles. This is very early US dinosaur literature.

He also finished a supplement to *Ichnology* in 1863, which was published in 1865 after his death.

The first book was illustrated with black-and-white lithographs while the supplement used the new and expensive technique of photographs pasted onto the pages and is only the third American scientific book to use photographs as illustrations.

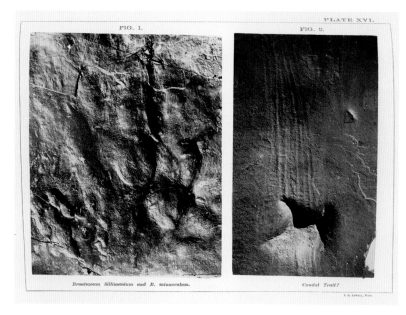

Julien Harmand
Guide élémentaire du lichénologue: accompagné de nombreuses espèces typiques en nature
1904
Height: 22 cm (8⅝ in)
Épinal

Julien Harmand (1844–1915) worked in the Nancy area of eastern France and was a priest as well as a well-known lichenologist. This guide, consisting of two albums of lichen specimens and a booklet, published in 1904 was aimed at amateurs.

The booklet sets out to define what lichens are, gives advice on how to study them, with and without a microscope and how to conserve them. He also lists the various classes and families of lichens.

The two albums contain 108 sheets of specimens with printed names. An unusual and rather charming work.

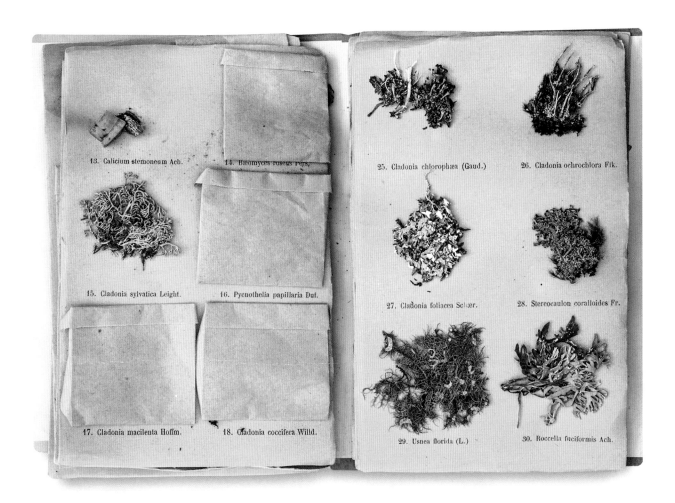

Dr JM Bourgery
Traité complet de l'anatomie de l'homme, comprenant la médecine opératoire, avec planches lithographiées d'après nature
Tome 6. Iconographie d'anatomie chirurgicale et de médecine opératoire
c.1831–54
2 vols. including lithography
by NH Jacob
Height: 45 cm (17¾ in)
Paris

Jean-Baptiste Marc Bourgery (1797–1849) was a French doctor who studied in Paris under Jean-Baptiste Lamarck (1744–1829), amongst others. In 1830 he set out to produce a complete treatise on human anatomy pertaining particularly to surgery. The final eighth volume was published posthumously in 1854. The collection includes both parts of the sixth volume of anatomical illustrations relevant to surgery.

The illustrations are masterly and were created by Nicolas Henri Jacob (1782–1871), a painter and draughtsman born into a family of artists and craftsmen, which included the famous cabinetmaker Georges Jacob. Nicolas was a pupil of the painter Jacques-Louis David (1748–1825) before he went on to become Professor of Drawing at the National Veterinary School of Alfort, just outside Paris. This school still exists and has a fine anatomy museum, which is well worth a visit. After his time at the Veterinary School he made the 700 lithographic plates for this publication and after Bourgery's death continued to work as a well-known scientific illustrator.

Bougery's teacher the naturalist Lamarck was, and remains, a very interesting figure. One of his major evolutionary theories was that characteristics acquired through the environment could be passed on – for example the giraffe that reaches up ever higher for food – a theory that Darwin's evolution by natural selection did not allow for. However recent developments in 'epigenetics' – the study of genetic control by factors other than DNA – suggest that external conditions can switch the expression of particular genes 'on and off' through chemical reactions, leading to heritable changes in individuals of a species.

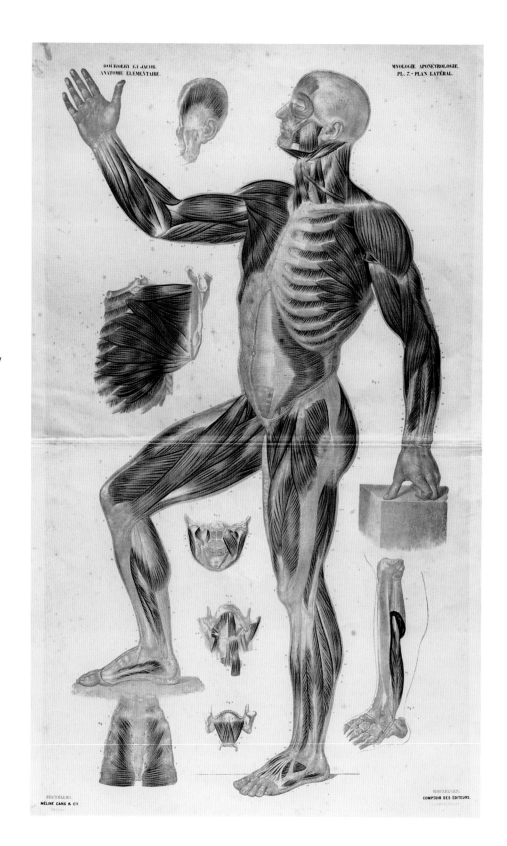

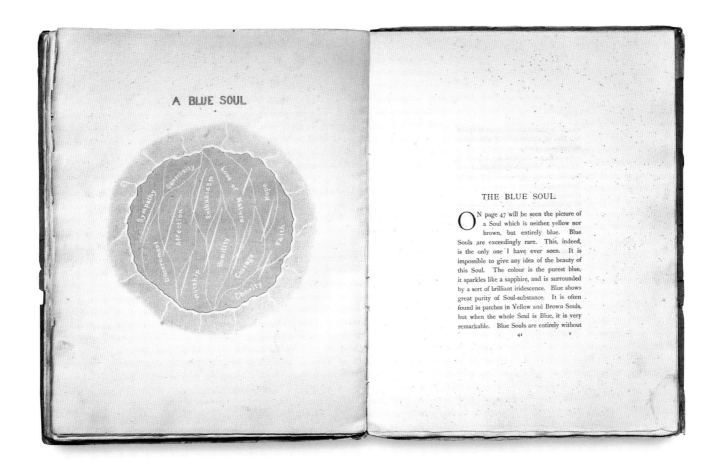

A BLUE SOUL

THE BLUE SOUL.

ON page 47 will be seen the picture of
a Soul which is neither yellow nor
brown, but entirely blue. Blue
Souls are exceedingly rare. This, indeed,
is the only one I have ever seen. It is
impossible to give any idea of the beauty of
this Soul. The colour is the purest blue,
it sparkles like a sapphire, and is surrounded
by a sort of brilliant iridescence. Blue shows
great purity of Soul-substance. It is often
found in patches in Yellow and Brown Souls,
but when the whole Soul is Blue, it is very
remarkable. Blue Souls are entirely without

41   F

Alice Murray Dew-Smith (?)
Soul Shapes
1890
4 hand-coloured plates
Height: 26 cm (10¼ in)
London

This is a very curious book. The author does not reveal him or herself, although it is reputed to have been a figure named Alice Murray Dew-Smith about whom very little is known. There is also a throwaway reference to Francis Galton, a Victorian polymath who was best known for his work in eugenics, although no further mention is made of Galton's theories.

The author states in the preface:

> My own peculiarity is seeing people's Souls in shapes and colours; and in the hope of provoking comparison I have made coloured diagrams of some of these mental images, and have added a short explanation of what the colours, shapes and individual peculiarities mean.

The illustrations are line drawings of varying shapes ranging from rough circles to amoeba-like shapes, with personality descriptions written into sections, and coloured following the author's code. They are what attracted me to this rather confusing book.

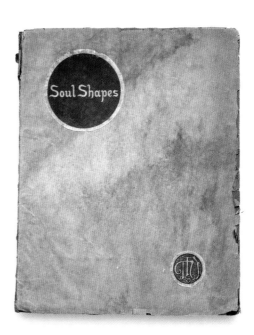

Dr Louis Thomas Jérôme
Auzoux
Papier-mache model of a
horse's hoof and fetlock
1893
Height: 30 cm (11¾ in)
France

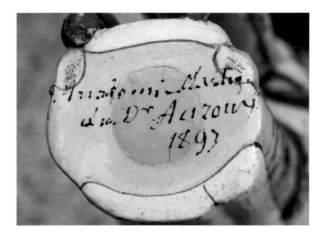

This very detailed model has a hinge at the front of
the hoof so that it can be taken apart for study, a
feature that all Auzoux anatomical models have in
common. The parts, such as muscles, vessels and
bones are numbered and would have been explained
in a booklet similar to that which accompanies the
turkey model in the collection (p.143).

It is very likely that this model was made for a
veterinary institute such as the École Nationale
Vétérinaire d'Alfort on the outskirts of Paris, which
has an outstanding collection of nineteenth-century
teaching material. It has also been suggested that
models like this one helped cavalry officers look
after their horses' hooves.

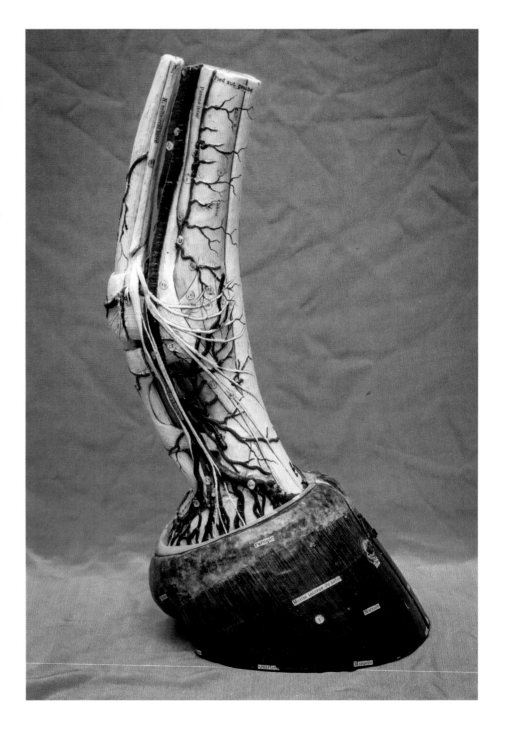

Richard Owen
Monographs of the British fossil reptilia from Oolitic formations
1861–81
1 vol. in 5 parts with illustrations by Joseph Drinkel, J Erxleben and A Gawan
Height: 29 cm (11⅜ in)
Palaeontographical Society, London

This is one of Richard Owen's classic works, produced over 20 years from 1861 to 1881. It deals with some of the well-known dinosaurs, the *Scelidosaurus* (allied to the *Iguanodon*), the *Plesiosaurus*, the *Pterosauria* (flying reptiles) and the *Ichthyosaurus*, which had all been the subject of considerable attention in France, Germany and the UK at that time.

The first three volumes were illustrated by Joseph Dinkel (1806–91), another of my favourite illustrators. Born in Austria, he became close friends with the Swiss naturalist Louis Agassiz (1807–73) in Munich in 1828 and made a large number of fossil fish drawings for him. Dinkel remained as Agassiz's draughtsman until the latter moved to Harvard University in 1848. In 1834 Dinkel moved to England where he also served as an illustrator for scientists such as Gideon Mantell (1790–1852) and Owen (see p.123).

Dinkel's most famous drawings of fish fossils are probably the ones for Agassiz's *Recherches sur les poissons fossiles* (1833–45; p.235).

Relatively little is known about J Erxleben and A Gawan, the illustrators of the last two volumes, other than that they were both very active illustrators.

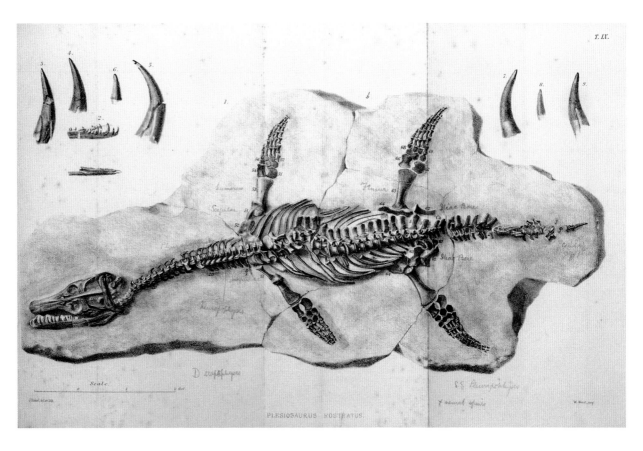

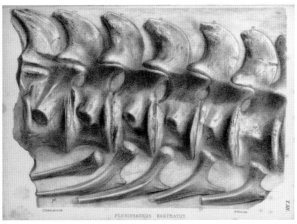

Richard Owen
Descriptive and illustrated catalogue of the fossil Reptilia of South Africa in the collection of the British Museum
1876
70 leaves of lithographs by C Griesbach
Height: 34 cm (13⅜ in)
London

This publication deals with fossil remains that were found in the late 1830s in the Cape Colony, a British colony in what is now South Africa, and subsequently sent to the British Museum where Richard Owen was Superintendent. At this point the Natural History Museum had not yet been established.

The largest part of the work deals with a group of dinosaurs named *Dicynodon*, meaning 'Two Dog Teeth', which were herbivores with prominent tusks, some 1.2 metres in length. Owen named a substantial number of species within this group.

The illustrations are by Carl Ludolf Griesbach (1847–1909), a palaeontologist and geologist who was educated in Vienna before joining a German scientific expedition to East Africa in 1869–70. He then lived in London for some years before he was appointed Assistant Superintendent to the Geological Survey of India, based in Calcutta, in 1878. He subsequently worked for the Afghan Boundary Commission and for the Amir of Kabul. He was also a very accomplished artist and his illustrations for this book are very beautiful indeed.

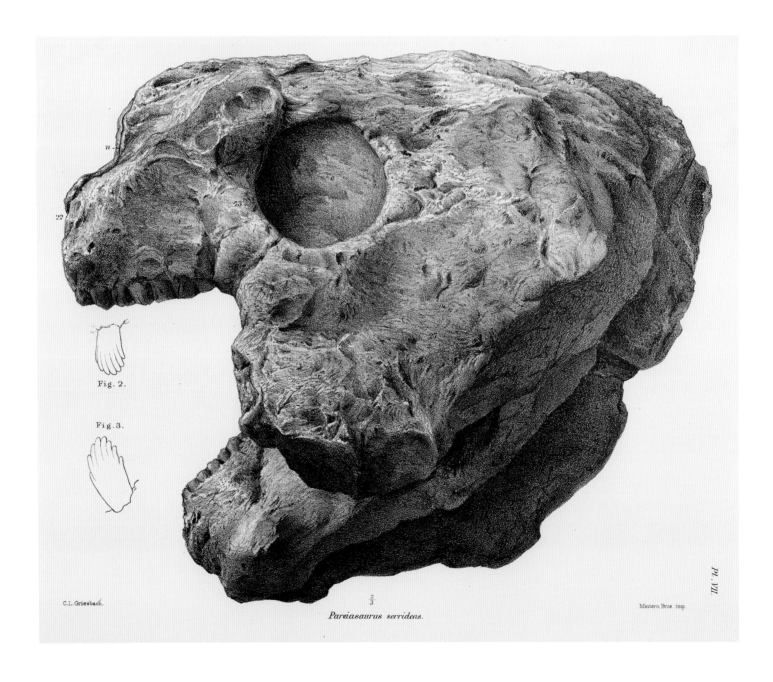

C.L.Griesbach.

Fig. 2.

Fig. 3.

Pareiasaurus serridens.

Mintern Bros. imp.

Pl. VII.

Leopold and Rudolf Blaschka
Three glass models of slugs
Mid-to late 19th century
Length: 9 cm (3½ in)
Germany

The Blaschkas (Leopold, 1822–95 and Rudolf, 1857–1939), were based in Dresden, and represent the holy grail of didactic nineteenth-century modelling. The Blaschka family business spanned around nine generations and lasted more than 300 years. Historically they worked in glass and made jewellery, test tubes and glass eyes before beginning to make scientific glass models in the mid-nineteenth century.

This kind of model was predominantly used by museums and universities as a method of demonstrating invertebrate species which were difficult to show in any other way. They ranged from jellyfish and octopi to radiolarians and ultimately to plants, which were made exclusively by the Blaschkas for Harvard University. Glass is, of course, ideal for showing translucent sea creatures; it is also malleable and can be painted or have colour added.

Although their pieces maintained a certain element of artistic freedom, scientific accuracy was always important. Both father and son went on field trips and read reports from scientific expeditions. Word of mouth and personal contact with naturalists led to a wide range of commissions that included producing models inspired by the work of their friend Professor Ernst Haeckel (see p.93).

In 1886 the Blaschkas were commissioned to make sample flower models for the Harvard University Botanical Museum and from 1890 these famous botanical models became their sole output. Leopold worked on the contract until his death in 1895 and Rudolf continued the work for the next 41 years, completing it just three years before his own death in 1936. He supplied the museum with over 4,000 replica flowers, representing 847 plant species, which are permanently on display at the Harvard Museum of Natural History, where they have many loyal admirers.

If you want to see how these models were used in nineteenth-century museums there is no better place than the Natural History Museum in Dublin – popularly known as 'The Dead Zoo' – an institution that has remained virtually unchanged since the end of the nineteenth century. The glass models there are still used as they were in Victorian times to show the various species of invertebrates in fairly full cases.

The majority of other museums generally ceased displaying Blaschka models in the middle of the twentieth century and replaced them with more modern display techniques such as photography. These models went into storage and were on the whole forgotten.

In 2002 an exhibition was organised at the Design Museum in London where Blaschka models were shown for the first time as works of beauty, rather than as scientific models. More recently, due to increasing interest in these models, pressure from Blaschka enthusiasts (including the Corning Museum of Glass in New York State) and a conference organised in Dublin in 2006, museums have hunted down their Blaschka models and have begun to display them as unique works of craftsmanship.

Blaschka models rarely appear on the market, as museums do not de-acquisition them. I learnt eventually that the ones in my collection all come from an English secondary school (which prefers not to be named) that closed down its natural history department for financial reasons.

These three slugs are the first Blaschkas I managed to acquire from a London dealer who was at first reluctant to say where he had got them or how many he had. However I didn't have long to wait before he sold me another one (see p.186) and I later exhausted his stock by buying up his last Blaschkas: ten models of slugs and snails and a wonderful Portuguese man o'war (p.227).

Richard Owen
Description of the skeleton of an extinct gigantic sloth, Mylodon robustus, Owen, with observations on the osteology, natural affinities, and probable habits of the megatherioid quadrupeds in general
1842
24 leaves of plates by George Scharf
Height: 32 cm (12⅝ in)
London

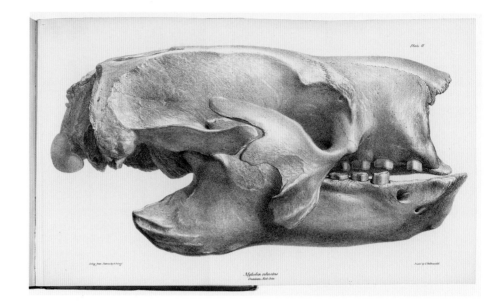

A fairly complete skeleton of a giant sloth was discovered near Buenos Aires, Argentina in 1841. It was purchased by the Royal College of Surgeons in London, along with the skeleton of a large armadillo-like creature, and added to a collection of other large, quadruped skeletons from Argentina donated by Charles Darwin in the late 1830s. Richard Owen, the naturalist and specialist in comparative anatomy (see p.123), studied all these bones.

George Scharf (1788–1860), my favourite illustrator besides Joseph Dinkel, made the illustrations. Scharf was educated in Munich and became an expert lithographer (see p.141).

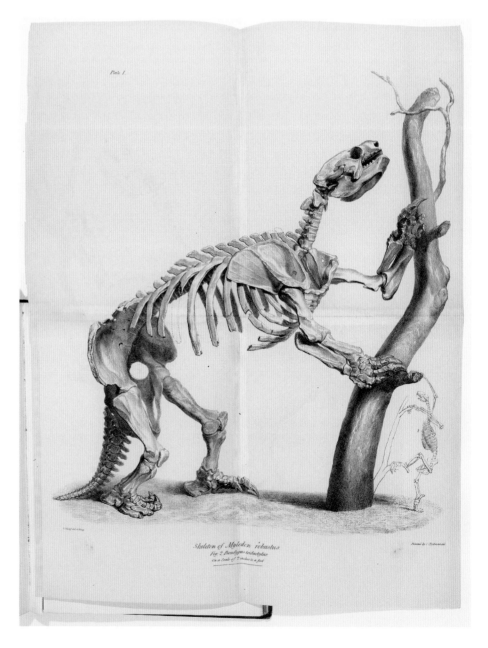

Thomas Wright
Monograph on the Lias
Ammonites of the British islands
1878–86
8 vols. with illustrations
by A Gawan and M Suft
Height: 28 cm (11⅛ in)
Palaeontographical Society,
London

Thomas Wright (1809–84) was a doctor who devoted himself to palaeontology and the collection of Jurassic fossils. He wrote 32 papers on geological subjects, the most important of which were for the Palaeontographical Society. His collection of fossils was ultimately sold to the Natural History Museum in London.

Ammonites are of course particularly attractive. They are an extinct group of marine invertebrate animals and are more closely related to living species such as the octopus, squid and cuttlefish than to the living Nautilus species. Geologists are particularly interested in ammonites because they can link rock layers in which a particular species is found to specific geological time periods.

These eight volumes are still in the original covers in which the printer delivered them.

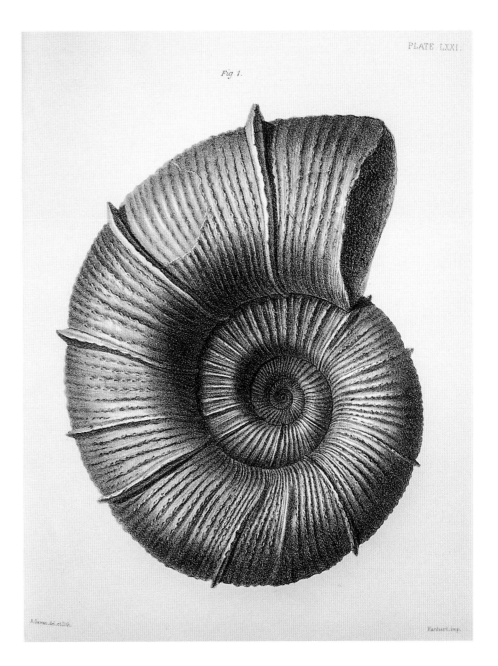

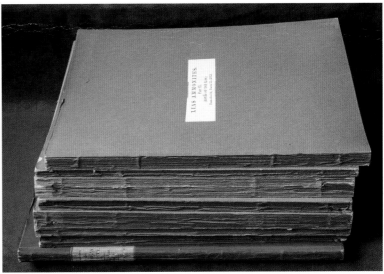

Fritz Kolbow
Wax model of a hand
Early 20th century
34 × 23 × 10 cm
(13⅜ × 9⅛ × 4 in)
Germany

This hand is quite intriguing: models of this kind tend to show some form of disease, however there is no obvious sign of this here.

The manufacturer is Fritz Kolbow (1873–1946), a German model maker who was active in Berlin from 1896. From 1910–22 he worked in Dresden, where the important Hygiene-Museum was being set up. He then returned to Berlin in 1922 and set up a business specialising in medical teaching material. His earlier work is well represented in the collections of the Charité Hospital in Berlin. Many similar wax models were recycled into candles in the 1960s.

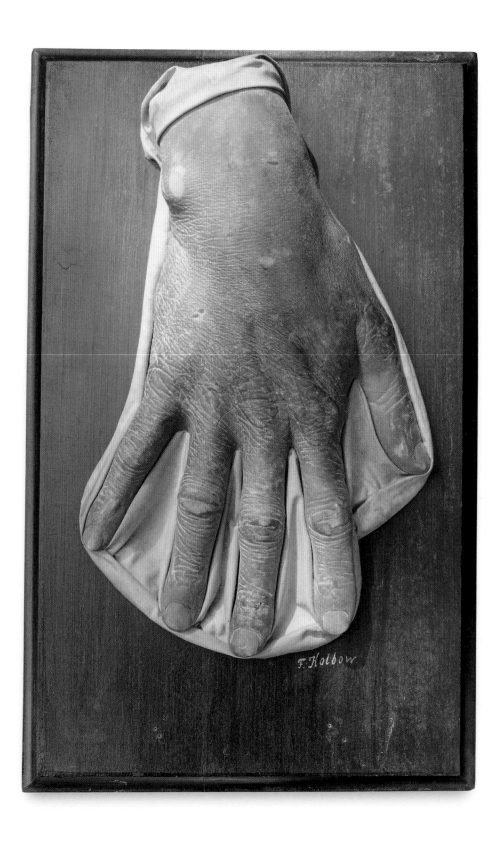

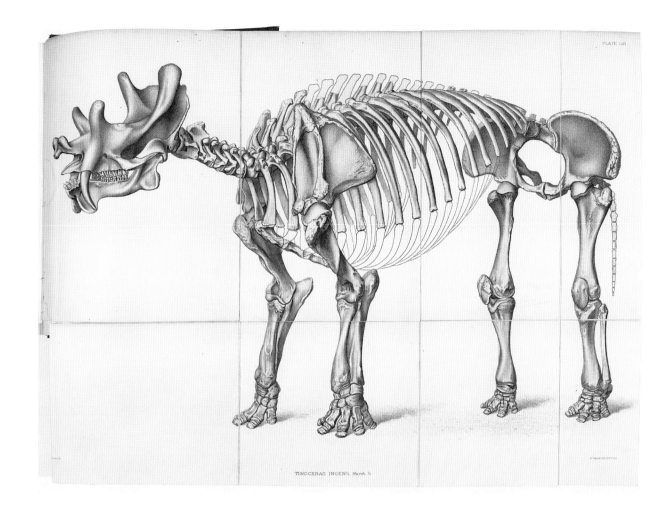

TINOCERAS INGENS, Marsh. ¼

PLATE LVI

Othniel Charles Marsh
*Dinocerata: a monograph
of an extinct order of gigantic
mammals*
1886
56 plates of illustrations
by F Berger
Height: 29 cm (11⅜ in)
Washington, DC

Othniel Charles Marsh (1831–99) was a great American dinosaur hunter, whose rivalry with his contemporary Edward Cope (1840–97) came to be known as the 'Bone Wars'. Using his uncle George Peabody's (1795–1869) extensive resources, Marsh collected for Yale University in New Haven, for what would ultimately become the Peabody Museum, while Edward Cope's collections ended up in the American Museum of Natural History in New York.

Marsh named many dinosaur species and published a number of his works under the auspices of the US Geological Survey (see p.100). The eastern slope of the Rocky Mountains provided the richest areas for finding dinosaur remains, but protection from the Native Americans had to be provided by the American Army. Prominent were the two civil war heroes Generals Sheridan and Sherman who were both heavily involved in the Indian Wars. Accounts of Marsh and Cope's expeditions make fascinating reading.

The illustrations in this publication are by F Berger. Judging by the large number of his unpublished illustrations in the Smithsonian's collection, Berger was a prolific illustrator.

Leopold and Rudolf
Blaschka
Glass model of a sea
anemone
Mid-to late 19th century
8 × 10 cm (3⅛ × 4 in)
Germany

This model of an anemone has clearly had quite an
active life. The plaster base is somewhat chipped
and it is mounted on a piece of cardboard, which
was added at a later date, and shows signs of
having been screwed down. Nevertheless it is a
typical Blaschka invertebrate, and uses techniques
that the father and son developed over time, such
as moulding the main body, attaching the finer
parts with glue and making the base of plaster.

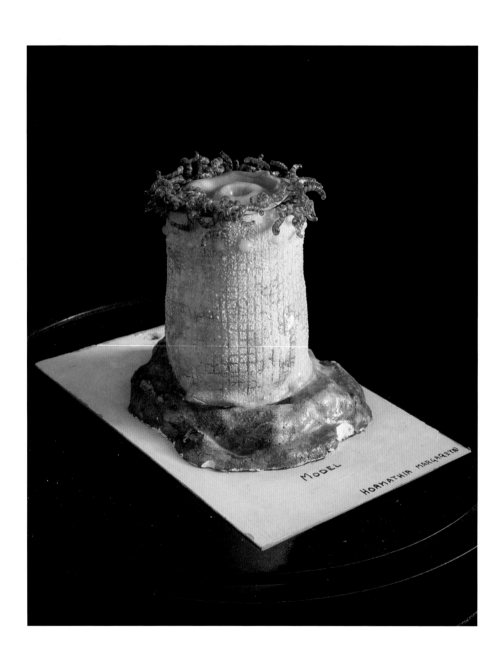

Dr HE Ziegler and
Dr Gerhard Kellar
Two steel models of
snake skulls
Late 19th century
Top: 55 × 22 × 4 cm
(21⅝ × 8⅝ × 1⅝ in)
Bottom: 54 × 26 × 1 cm
(21¼ × 10¼ × ⅜ in)
Germany

These thin flat steel models of snake skulls depict the jaw mechanisms of the *Python sebae* (African rock python) and the *Crotalus durissus* (South American rattlesnake). A label on the back states that the models were made under the instruction of Dr HE Ziegler and Dr Gerhard Kellar of Stuttgart. Little is recorded about either, although this Ziegler is not the Ziegler (Adolf) who made hundreds of wax models over his career (see p.189).

An elaborate text on the back of the models describes the connections and possible disconnections between the jaw and the skull that enables the snake to swallow very large prey.

The Hunterian Museum at Glasgow University has a similar model of the rattlesnake. At first glance they look like drawings because they are so thin but they are, in fact, elaborately crafted, nearly mechanical instruments.

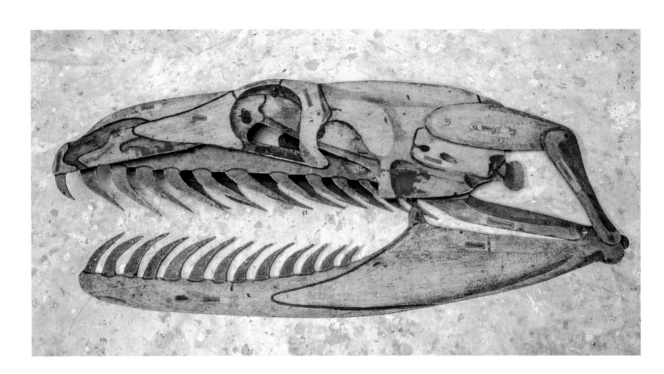

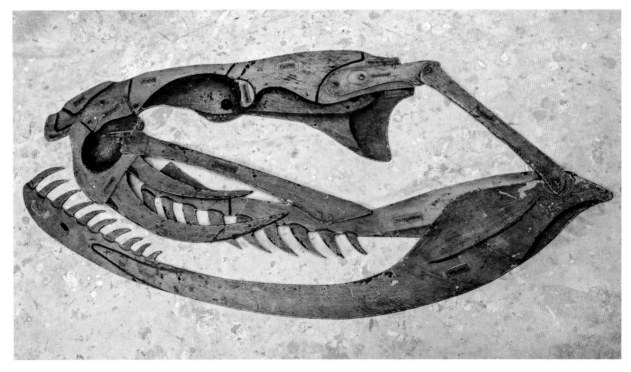

25 ivory models of fruit
and vegetables
19th century
Each, length between:
4–11 cm (1⅝–4⅜ in)
India

It is not clear what these models were created for
but what is certain is that they were produced in
India and are made of coloured ivory. A few of the
models are hollow and have stems that unscrew,
which suggests that they could be used to hold
some kind of fine material.

India has a tradition of making models of fruits
and vegetables for a variety of purposes. This
grouping seems very handsome, whatever its
purpose, and I included it in the first part of the
book (p.23) as it looks somewhat like a painting.
This item is another in the collection that
demonstrates how the viewer or collector can
attribute a new meaning to an object the original
purpose of which has been lost or forgotten.

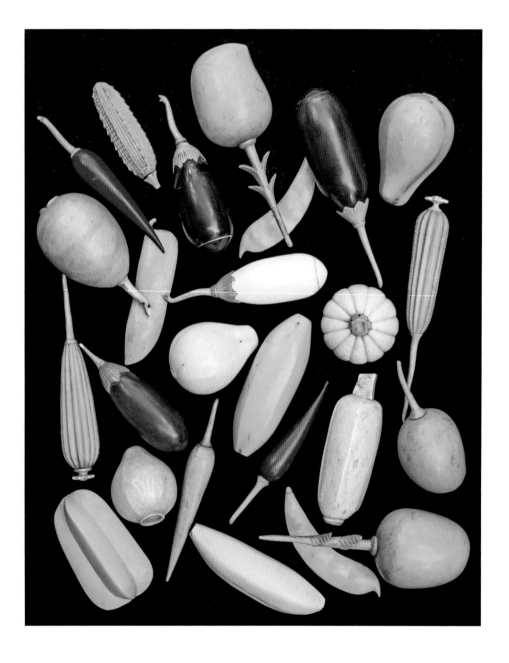

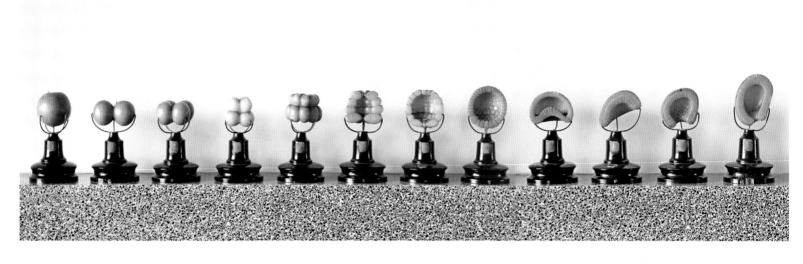

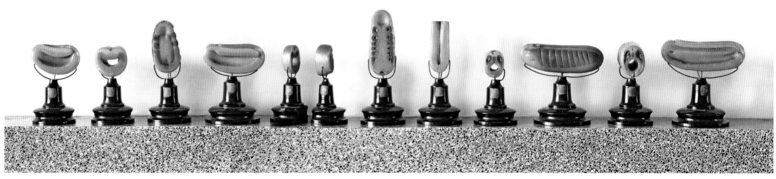

Adolf Ziegler
25 wax models of a
developing *Amphioxus*
After 1882
Each, between:
11–17 × 6–9 cm
(4 ⅜–6 ¾ × 2 ⅜– 3 ½ in)
Germany

This is a series of the development of the embryo of the *Branchiostama* (part of the order of *Amphioxoformes*), which is one of the few living examples of lancelets, a small fish-like creature measuring 5–7 centimetres that lives half-buried in the sand in shallow temperate and tropical seas. It was considered of scientific interest because it is a close relative of early vertebrates, which split off some 520 million years ago.

This archetypal vertebrate form has no skeleton but, in common with vertebrates, does have a hollow nerve cord, which is protected by a cylinder of cells. This similarity to early vertebrates led researchers to study its embryo.

From the models, you can see that only in the eighth stage does the splitting of the cells lead to a differentiation of cell types (the cells are colour-coded when they adopt more specific functions such as external or various internal parts). Later on, you can see the neural hollow developing.

Adolf Ziegler (1820–89) was a medical doctor and laboratory assistant, who moved from Mannheim in Germany to Freiburg, where he made evermore

sophisticated models in multiple copies. With the help of microscopes, he made hard wax or clay moulds from which these series were then made. By the 1860s, he was selling internationally. Ziegler's son lived until well after the First World War, when Freiburg was damaged and his business reduced. His company was later taken over by the Sommer family in Sonneberg, the toy-making capital of Germany. Later this territory would become East Germany at which time the company was expropriated. The Sommer family moved to West Germany and continued making scientific models, mainly in plastic.

A discussion of these models would not be complete without mentioning Dr Nick Hopwood, Senior Lecturer at the Department of History and Philosophy of Science, University of Cambridge, whose publication *Embryos in Wax* (2002) is the definitive text on Ziegler and reflects his enthusiasm for these models.

This Adolf Ziegler should not be confused with Hitler's favourite artist with the same name, who lived from 1892 to 1959.

Dissected human skull
Early 20th century
15 × 12 × 22 cm
(5⅞ × 4¾ × 8⅝ in)
Europe

This item is not particularly old but the way in which it has been deconstructed and has had certain bits removed is quite peculiar. The object has been put together so that it can be opened up in various places: the top of the skull – the inside of which is painted with blood vessels – can be removed; the jaw is attached to a set of springs; part of the upper jaw can be removed to show the cheekbone and some teeth, the lower jaw on the right hand side can be opened to reveal the teeth that have been bisected (a difficult operation) and their roots; and a flap above the left eye opens up to reveal a cavity and the texture of the bone. Coloured threads around the jaw show vessels and nerves leading to the teeth.

The extreme craftsmanship and the bleaching of the skull has produced something that is far removed from its origin. It has been completely depersonalised (see pp.91, 135).

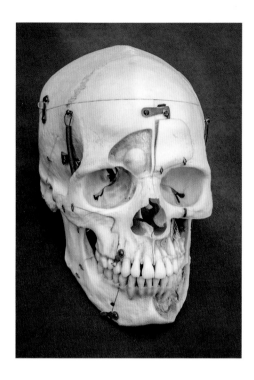

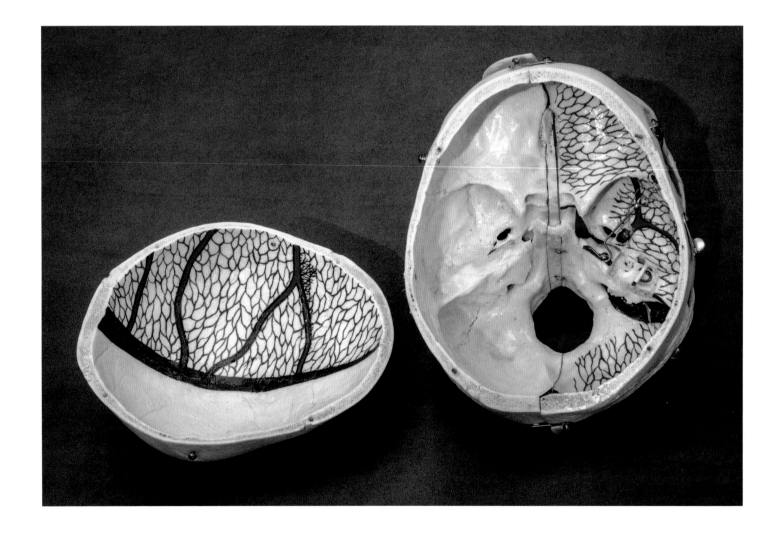

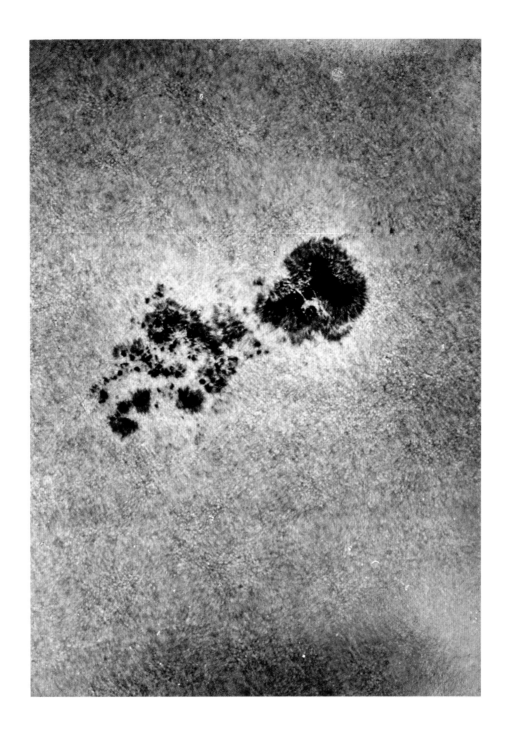

J Janssen
Annales de l'Observatoire
d'astronomie physique de Paris,
sis Parc de Meudon
(Seine-et-Oise)
1896
9 heliogravures and 12
photographic studies of the
solar surface by P Dujardin
mounted on card with
printed captions
Height: 29 cm (11⅜ in)
Paris

Jules Janssen (1824–1907) was a French astronomer who used spectroscopy – the analysis of the spectrum of visible and invisible light – to conduct his observations outside the earth's atmosphere. These are remarkable photographic studies of the solar surface (see also p.2) taken in the late nineteenth century. Janssen states that 'the photographic plate is the retina of the scientist'.

Janssen's research regarding very low frequency rays made the chemical analysis of planetary atmospheres possible. He was also able to

announce the presence of water vapour in the atmosphere of Mars as early as 1867, an observation that is still controversial.

Today the Parc de Meudon is the principal location of the Paris Observatory, which was originally founded in the Montparnasse district of Paris a few years before Greenwich was built in 1675. The original Chateau de Meudon was destroyed in the Franco-Prussian War and rebuilt in 1874 as a more up-to-date facility to which the Paris Observatory moved most of its activities.

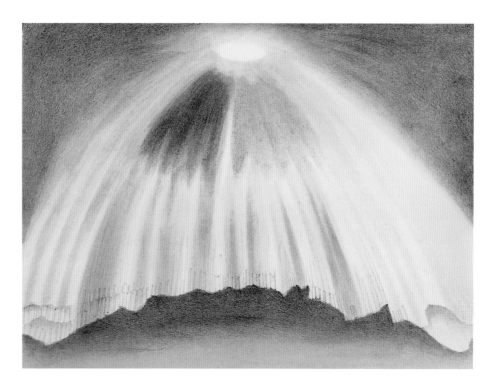

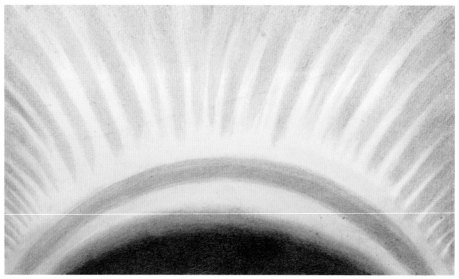

John H Morgan and
John T Barber
*An account of the aurora
borealis, seen near Cambridge,
October the 24th, 1847: together
with those of September 21, 1846,
and March 19, 1847, seen at the
Cambridge Observatory: with
twelve coloured engravings*
1848
12 leaves of coloured
engravings by J Andrews
Height: 27 cm (10⅝ in)
Cambridge

This is a very elaborate description of what took place on 24 October 1847 in Cambridge, when a spectacular manifestation of the Northern Lights was seen. The Royal Observatory in London had recorded significant magnetic disturbances and changes in the measurement of atmospheric electricity just before the lights were visible. However it was not until much later, towards the end of the nineteenth century, that it was discovered that what is now well-known as the aurora borealis is produced by charged particles from sun spots affected by lines of magnetic force colliding with the Earth's atmosphere.

The 12 very fine illustrations are by the artist James Andrews (1801–76), who is better known for his flower illustrations. He used the zincography process, where limestone is replaced with zinc plates, which were easier to use and cheaper than the limestone lithography.

The images appeal to me as almost abstract art.

This is a sponge of the kind found in the deep
ocean in the Western Pacific, particularly in the
area between Japan and the Philippines. It lives
symbiotically with two small shrimp inside it.
The body of the sponge is composed of silica,
which has the appearance of being 'woven' into
a fine mesh. What you see here is the skeleton
of the sponge that has been bleached to produce
the pale white colour.

When you compare a real object like this sponge
with the models that the Blaschkas created
you can see just how remarkable their work is
(see pp.181, 186, 227).

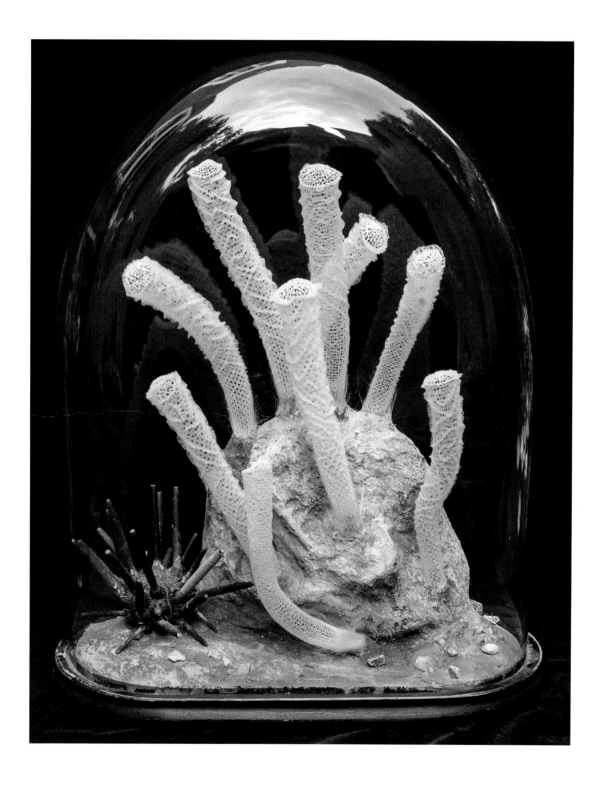

Deyrolle
Three plaster plaques
of dissected squid
19th century
Left: 53 × 35 × 4 cm
(20⅞ × 13¾ × 1⅝ in)
Right and not illustrated:
53 × 40 × 4 cm
(20⅞ × 15¼ × 1⅝ in)
Paris

These squid on wooden boards are made of painted plaster, probably in the late nineteenth century by the famous Paris establishment of Deyrolle. Founded in 1831, Deyrolle manufactured and traded in didactic natural history material, such as insects, butterflies and general taxidermy. They later expanded into teaching aids for schools and exported their wares to some 120 countries.

Their shop has been in the rue du Bac in Paris since 1888. Although a fire destroyed most of their stock in 2008, the building survived and, after a successful fundraising effort, Deyrolle started again and is currently doing well.

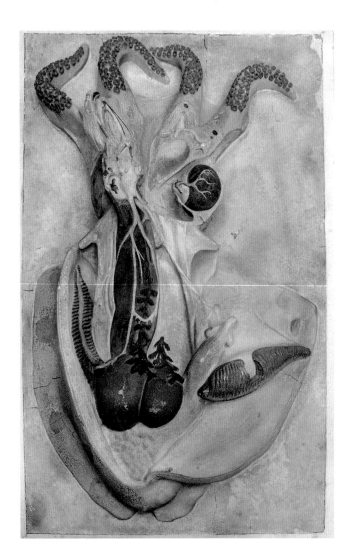
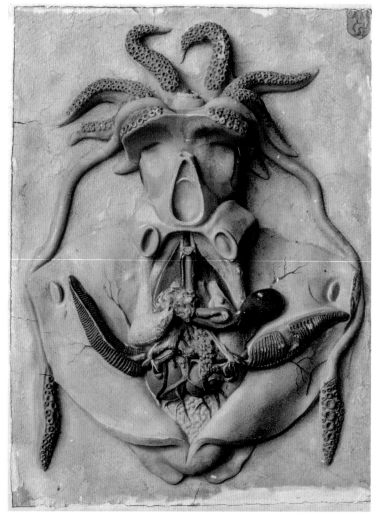

Thomas Moore
*The ferns of Great Britain
and Ireland*
1855
51 leaves of colour plates
by Henry Bradbury
Height: 57 cm (22½ in)
London

*Nature-printed British ferns:
being figures and descriptions
of the species and varieties
of ferns found in the United
Kingdom*
1863
165 leaves of colour plates
by Henry Bradbury
2 vols.
Height: 25cm (9⅞ in)
London

These are two books dealing with the same topic,
namely British ferns, by the same author and
nature-printer. One is large, folio size, the other
standard, octavo size. The folio edition is one of the
great nature-printed books. It was the first book
in the United Kingdom to use the new process of
nature-printing (see pp.81, 107) and includes 51 hand-
coloured plates.

The smaller, two-volume octavo edition has in
the first part colour plates of the same species as
the folio edition and adds another 114 new species
in the second part. Although it was produced eight
years later than the folio edition, the plates would
have been produced before the nature-printer
Henry Bradbury's death in 1860.

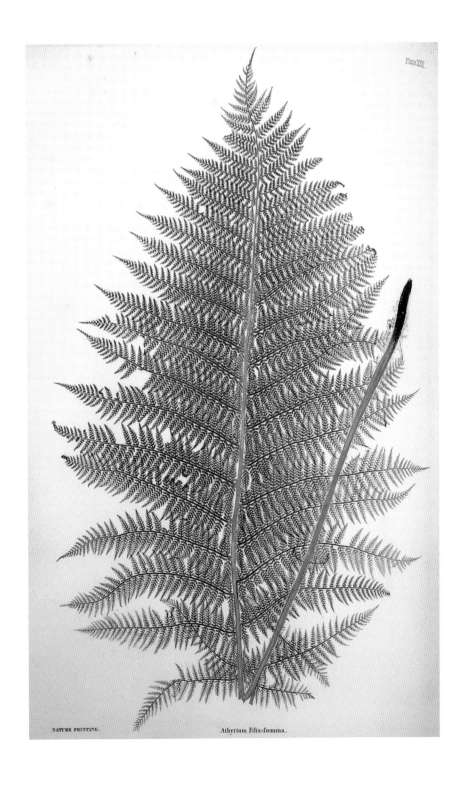

NATURE PRINTING.   Athyrium Filix-foemina.

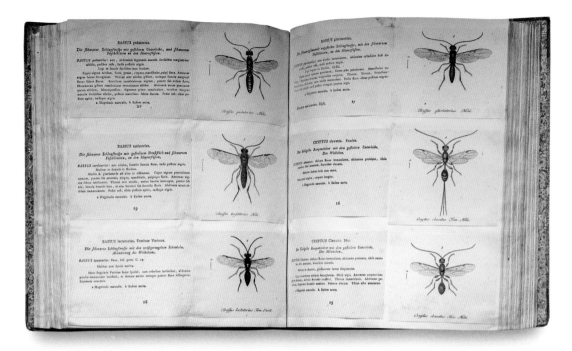

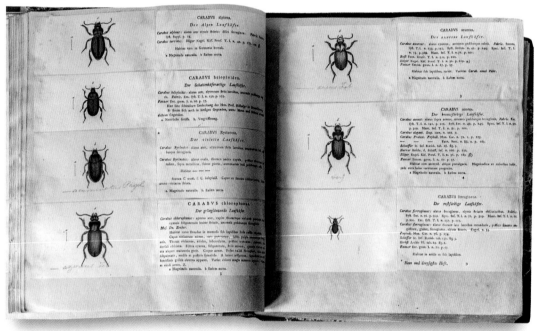

Georg Wolfgang Franz
Panzer
*Faunae insectorum germanicae
initia: oder Deutschlands
Insecten*
1793–1808
2,420 mounted hand-
coloured plates
7 vols.
Height: 27 cm (10⅝ in)
Nuremberg

Georg Wolfgang Franz Panzer (1755–1829) was a
German doctor and a celebrated botanist and
entomologist. The collection includes seven
volumes of his *magnum opus* on German insects
which features some 2,500 hand-coloured etchings
with accompanying printed text stuck onto
individual pages. The complete work was issued in
109 parts between 1793 and 1808.

The artist for this publication, Jacob Sturm (1771–
1848), was an internationally famous entomologist

and engraver of natural history illustrations, based
in Nuremberg.

During this period Nuremberg was the centre
of natural history book production in Germany. This
stretched back to the seventeenth century when
the combination of Nuremberg's status as a free
imperial city with no local ruler and its favourable
location on the crossroads of various trading routes,
led to its emergence as the northern European
centre of humanist learning and book printing.

Carl Gustav von Carus
*Von den Ur-theilen des Knochen-
und Schalengerüstes: mit XII
Kupfertafeln und einer
schematischen Schrifttafel*
1828
12 leaves of plates
Height: 43 cm (16⅞ in)
Leipzig

Carl Gustav Carus (1789–1869) was a doctor of medicine and a polymath connected to the German Romantic movement. He was also a friend of Johann Wolfgang von Goethe (1749–1832).

This work is an early, very scientific treatise on the archetypal bone structures of vertebrates and shellfish. Carus, who had studied painting under the German Romantic painter Caspar David Friedrich (1774–1840), etched the illustrations himself. Carus is best known for postulating the concept of the vertebrate archetype, an important concept in Darwin's ideas on evolution.

The illustrations in this work are dissections of bone and skeletal structures that take the form of exploded diagrams possessing an almost abstract quality, more akin to engineering diagrams than to anatomical drawings, which is why I like them so much.

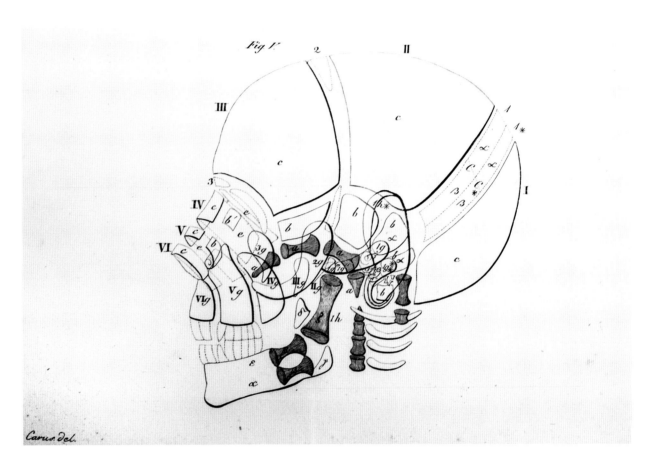

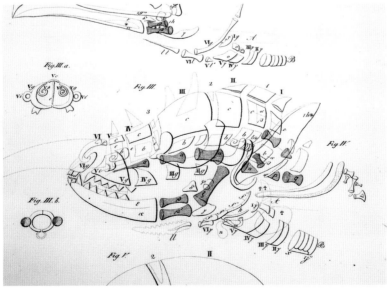

Charles Landseer
Écorché drawing
of the left hand
Mid-19th century
38 × 54 cm (15 × 21¼ in)
England

Charles Landseer (1799–1879) was a member of a family of artists. His brother Edwin (1802–73), largely known as a painter of animals, was the more renowned. Charles was trained at the Royal Academy of Art in London, where the Anglo-Swiss artist Henry Fuseli (1741– 1825) was an inspired teacher. He was also taught by the colourful figure Benjamin Robert Haydon (1786–1846). Although he was reasonably successful as an artist during his lifetime, Haydon was controversial and alienated his supporters, which led to regular financial trouble and ultimately his suicide.

In the debate between the view that in art 'the knife should go with the pencil' (i.e. study nature) versus the argument that artists should study the classics, Haydon opted for dissection to learn how to represent the human body.

Anatomy had been actively taught at the Royal Academy of Art by William Hunter (1718–83), the famous surgeon and anatomist. After his death in 1783 anatomical instruction at the Royal Academy declined and Haydon and some of the other students took anatomy classes from Charles Bell (1774–1842), a well-known surgeon and anatomist, whose anatomy theatre was in Soho, London.

A contemporary report states that: 'Thomas [another brother] and Charles Landseer have been concluding their preparatory studies as far as concerns dissection. Three weeks they have been hanging over a putrid carcass, dissecting and drawing for 12 and 15 hours a day at a time of the year when surgeons generally give up. [In summer the bodies smelled.] They have made some capital drawings…'

In this drawing Landseer has used the *écorché* technique – a figure drawn showing the muscles of the body without the skin – to show the wrist and fingers of the left hand. It shows the 'extensors': the muscles that extend or straighten a limb or body part and these and some other tissues are numbered on the drawing and listed on the left side of the paper.

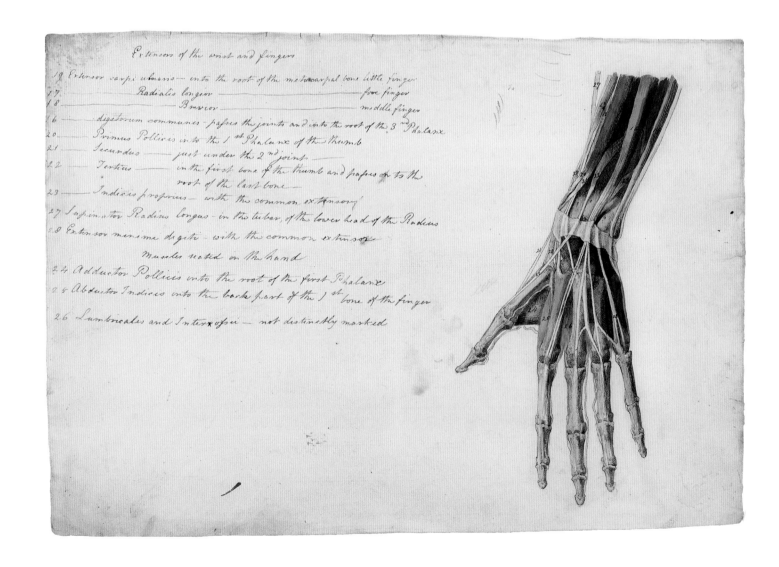

Václav Frič
Six plaster models of fish
Late 19th century
Each, between:
36–76 × 12–23 × 7–17 cm
(14⅛–29⅞ × 4¾–9⅛ × 2¾–6¾ in)
Prague

These model fish, complete in their original boxes, would have been used for teaching purposes. Frič was the well-known dealer in natural history educational material in Prague, Czech Republic (see p.148).

The original skins of the fish were preserved and were then used to cover moulded bodies made of plaster, which is why they appear so realistic. The boxes look like they come from an old-fashioned glove shop.

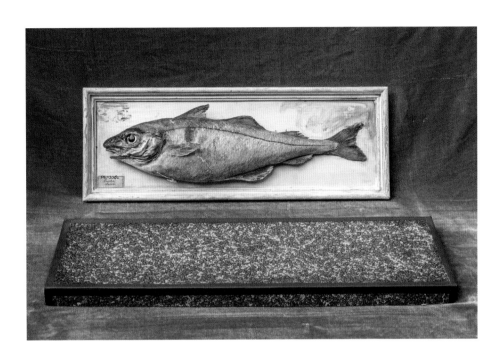

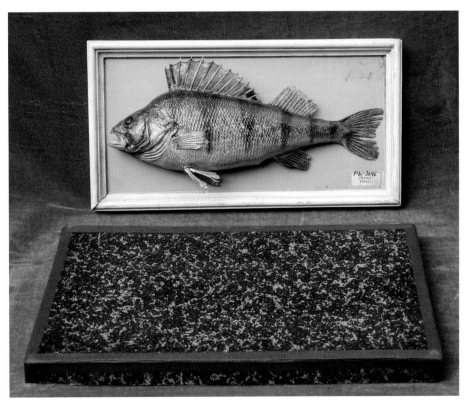

Carved ivory banana
1868–1912 (Meiji Era)
13 × 5 × 5 cm (5⅛ × 2 × 2 in)
Japan

This small replica is clearly not made for scientific purposes. It is part of the Japanese tradition of carved ivory items and has no didactic content, other than the extremely fine detail with which part of the skin is peeled away, showing the flesh inside.

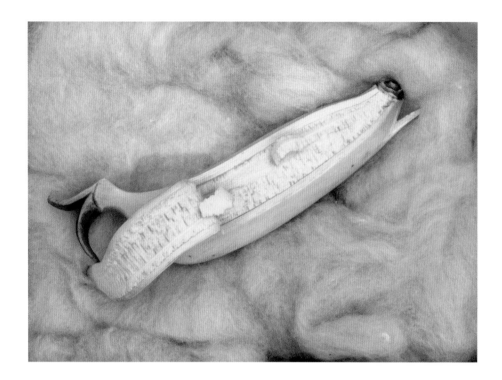

Chikusai Kato
Nine botanical paintings on
wood panels
c.1878
Each: 34 × 22 × 3 cm
(13 ⅜ × 8 ⅝ × 1 ⅛ in)
Tokyo

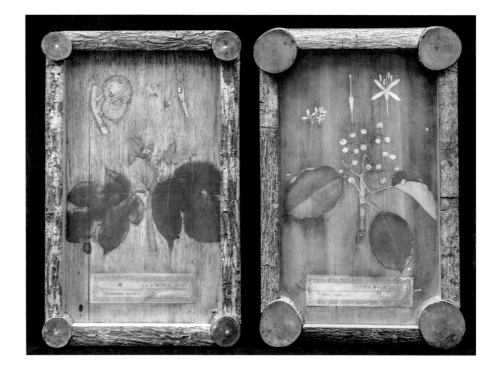

These Japanese panels were made for the comparative study of trees and for public display. They are painted illustrations on board cut from the wood of the species being shown. The 'frames' are from the bark of that particular tree. The panels are all labelled with Latin (Linnaean) binomial names as well as the common Japanese names.

In Europe during the eighteenth and nineteenth centuries, collections of wood samples were quite widely used for teaching, although today few remain on display. Philipp Franz von Siebold, the doctor at the nineteenth-century Dutch trading base on the island of Deshima in Nagasaki harbour (see p.119), brought back a beautiful collection of Japanese wood samples, which is now at Naturalis Biodiversity Center in Leiden, The Netherlands.

These painted panels were made c.1878, quite soon after the opening up of Japan in 1867, either by or under the supervision of Chikusai Kato (1813–?), the first botanical artist at Tokyo University.

The Botanical Museum Berlin-Dahlem has 152 in its collection, which is the largest set of these panels. Kew Gardens has 26, Harvard holds another eight and Tokyo University has 25. Together with my nine panels this totals 220, representing 150 different species of Japanese plants. It is believed that these panels soon became so desirable that in the West they were dispersed over several collections. One of my panels has a Japanese auction sticker on the back.

I obtained these objects through a dealer in Vienna, who had found them in The Netherlands.

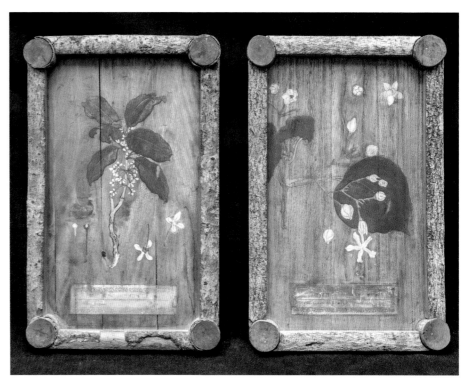

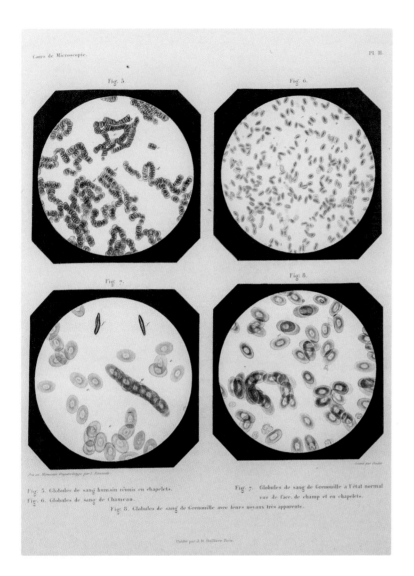

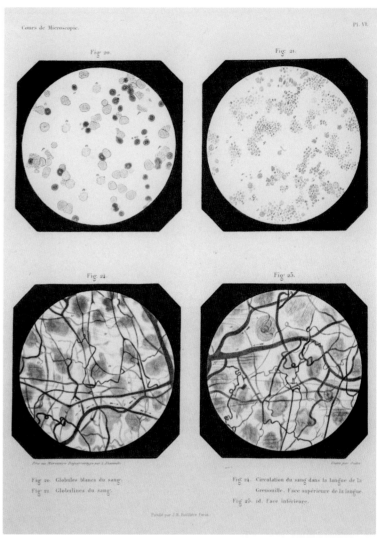

Alfred Donné and Léon
Foucault
*Cours de microscopie:
complémentaire des études
médicales: anatomie
microscopique et physiologie
des fluides de l'economie:
Atlas executé d'après nature au
microscope-daguerrotype*
1845
20 leaves of plates:
microscope daguerreotypes
by Léon Foucault
Height: 42 cm (16½ in)
Paris

Alfred Donné (1801–78) was a doctor and bacteriologist who discovered, amongst other things, leukaemia and also invented the photoelectric microscope.

He was assisted in his microscopic and anatomical work by Léon Foucault (1819–68) who started studying medicine but moved into physics, as he could not stand the sight of blood. Foucault interested himself in the daguerreotype process of photography and became an assistant to Donné. Later he became more famous than Donné, after he was involved in measuring the speed of light and demonstrating the rotation of the Earth, for which he constructed his famous pendulum, still to be seen in the Panthéon in Paris.

This work was produced for a course in microscopy as part of the study of medicine, particularly of microscopic anatomy and the physiology of human bodily fluids. The illustrations are based on microscope daguerreotypes subsequently engraved onto plates. They show various liquids such as blood, mucus, mother's milk, pus, saliva, urine and sperm, as well as crystals and eggs.

The images illustrated here show blood cells of a camel, three of a frog, and four of humans.

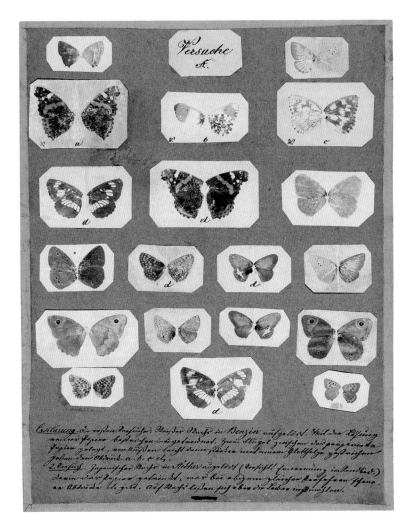

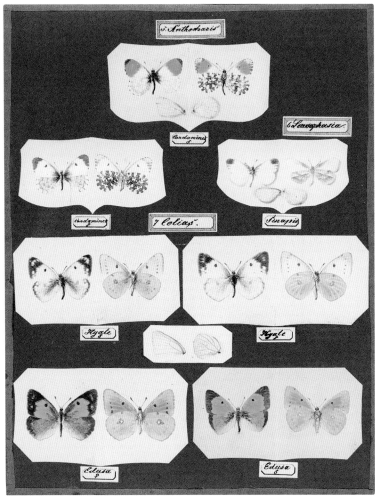

Butterfly wing print trials
Mid-19th century
Each: 33 × 26 cm (13 × 10¼ in)
Germany

In these attempts at nature-printing the anonymous makers of the 12 sheets have tried to make prints directly from butterfly wings.

The idea was that the pigments in the wings would be transferred to the paper, a method which of course meant that it was not possible to make more than one or two prints.

The technique involved soaking the butterfly wings in a mixture of benzene and melted wax or 'Japanese' wax and ether, and then pressing the wings hard onto paper. The makers state that the experiment was not satisfactory. Later this technique became more successful, as demonstrated by an album of Mexican butterflies from 1862–67 printed in this way (p.223). In 1900 a book was published in Boston, entitled *Moths and Butterflies of the United States, East of the Rocky Mountains* by Sherman F Denton which displayed butterflies printed in the same manner. It was published in an edition of 500 and the author states having used 50,000 specimens to produce the illustrations.

*Dendrophyllia, Gorgonian and
Madrepore* coral specimens
19th century
Left: 19 × 9 × 9 cm
(7½ × 3½ × 3½ in)
Centre: 20 × 9 × 9 cm
(7⅞ × 3½ × 3½ in)
Right: 36 × 9 × 9 cm
(14⅛ × 3½ × 3½ in)
Europe

These three coral specimens, *Dendrophyllia*, *Gorgonian*
and *Madrepore* are typical late nineteenth-century
teaching items. They are all properly scientifically
named and clearly come from the same source,
as the bases are identical and two of them have
similar labels, which state '885' and '878'.

Other than this, little is known about the provenance or manufacture of these specimens. They
have a certain rugged beauty.

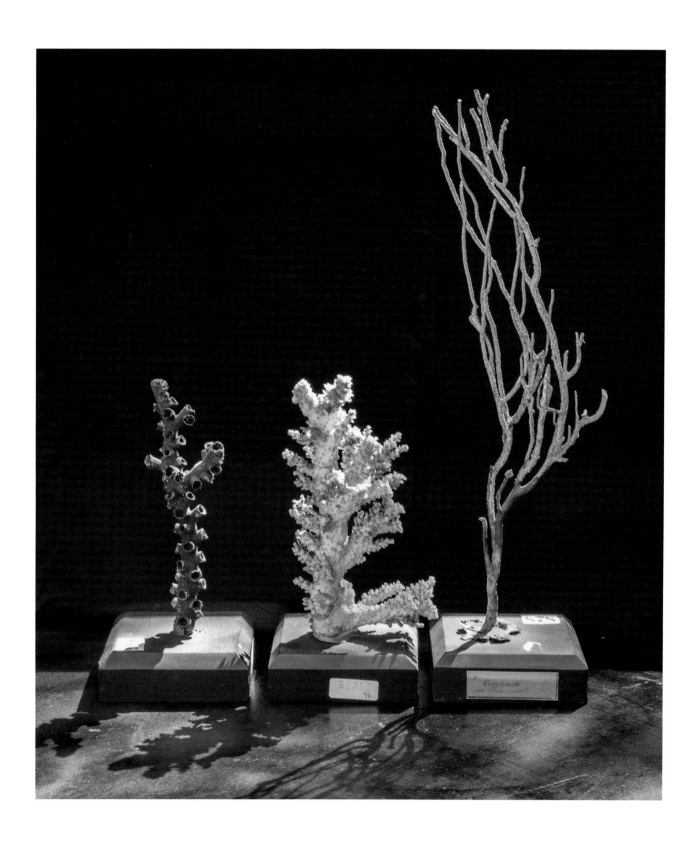

Taxidermied toad
19th century
8 × 18 × 11 cm
(3⅛ × 7⅛ × 4⅜ in)
Europe

I am not usually particularly interested in taxidermy, however this toad has a certain presence and my wife Angie is very fond of it: 'I find toads majestic in their ugliness.'

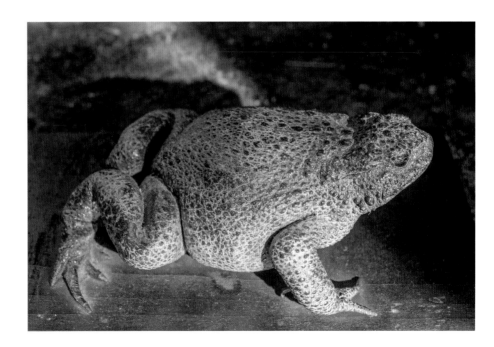

Max Ernst Wichura
Boxed herbarium of
Lapland flora
1860
11 × 26 × 16 cm
(4⅜ × 10¼ × 6¼ in)
Lapland

This boxed collection of botanical samples from Lapland includes 100 herbarium sheets of flowering plant samples together with 30 sheets of seaweeds and mosses. Also included are 74 pen and ink caricatures, including a number that depict botanists collecting samples.

Max Ernst Wichura (1817–66) was a German lawyer and botanist. He was a government official but also found time to go on plant hunting expeditions, such as this one to Lapland in 1856. In 1859 he was part of a Prussian expedition to the Far East and visited Singapore, Manila and the coastal areas of China and Japan. He also corresponded with Charles Darwin.

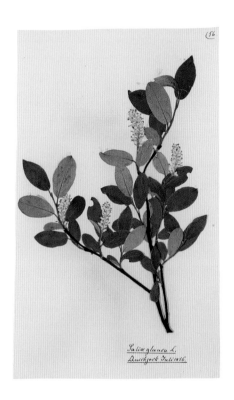
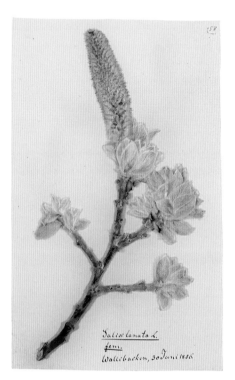
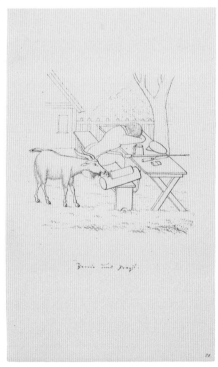
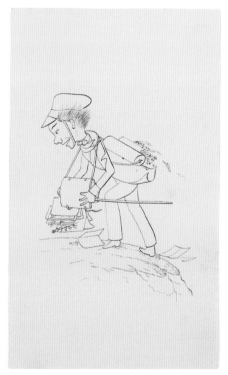

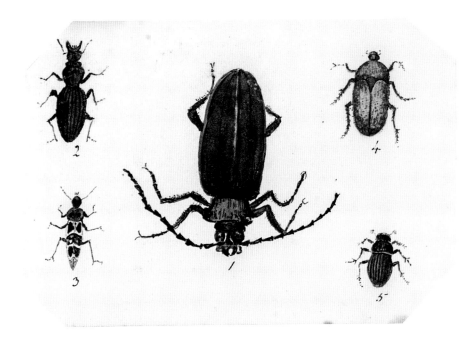

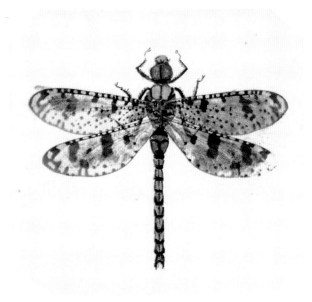

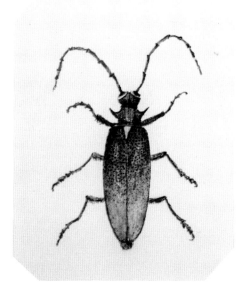

Album of butterfly and insect specimens
1859–64
30 × 20 × 2 cm
(11¾ × 7⅞ × ¾ in)
India

This charming album, which looks a bit like an old-fashioned school exercise book, is a collection of insects made by someone who lived in the hill station Mount Abu in northwest India during the middle of the nineteenth century.

The specimens consist mainly of wings stuck onto pieces of paper, with bodies coloured in, and inserted into the individual pages of the album. The collector has identified each insect and indicated where it was found. Many of the specimens were found in a location referred to as the 'school compound' – including larger insects, which might have given the school children a bit of a turn.

Mount Abu belonged to the ruler of Rajputana who leased it to the British early in the nineteenth century. The cooler hill stations were important to the British to recover from the heat and diseases of the tropics. Gradually hill stations such as Mount Abu expanded into larger communities.

Mount Abu still exists as a hill station at 1,700 metres above sea level and is popular with Indian honeymooners.

Eight wax models of
carnations
19th century
Each, between:
38–46 × 14 cm (15–18 × 14 in)
Tuscany

Fine wax models from the late eighteenth and early nineteenth centuries can be found in Florence, Italy. The Grand Duke of Tuscany Pietro Leopoldo (1747–92), the younger son of Maria Theresa, Empress of Austria (1740–80), embarked on a policy of enlightenment. This brief window, before the upheavals of the French Revolution and Napoleonic Wars, coincided with something similar in Spain, which also allowed a short flowering of scientific enquiry and included expeditions, such as Alexander von Humboldt's to South America in 1799. Pietro Leopoldo's rule ended in 1790 when he left Tuscany for Vienna to succeed his brother Joseph II. He died soon after in 1792.

One of Leopoldo's lasting legacies was the foundation of the Museum of Natural History in 1775 and the Astronomical Observatory (La Specola) in 1790. Now merged, they are still today located behind the Pitti Palace in Florence. Until well into the nineteenth century, the Specola was one of the few scientific museums open to the public. Today it still exhibits life-sized wax anatomical models, including cutaway models of pregnant women, clearly made in the same tradition as church statuary. The women have the look of the Madonna. Though they lie in a glass case with their innards on display, their elegant legs are chastely crossed at the ankles, their demure faces turned away from the onlooker, and some wear a pearl necklace. Apart from the famous anatomical models, the museum also had wax botanical models, which are now in the university there.

The flowers shown here, as well as various waxed fruit (pp.220, 237), were all acquired from the same source in Northern Italy. They were made in the mid-nineteenth century for an agricultural or horticultural teaching institute, as part of a push into more scientific farming.

The carnations are very beautifully made but, like the wax fruits, their labels are more recent. The pots are of earthenware and are – at least partially – filled with lead in order to stabilise the whole. Due to their delicacy and weight, their teaching function was clearly more restricted than the papier-mache teaching models, which were made to be handled and taken apart (p.105). These flowers are included in the first part of this volume (pp.40–41) because of their hyperrealism.

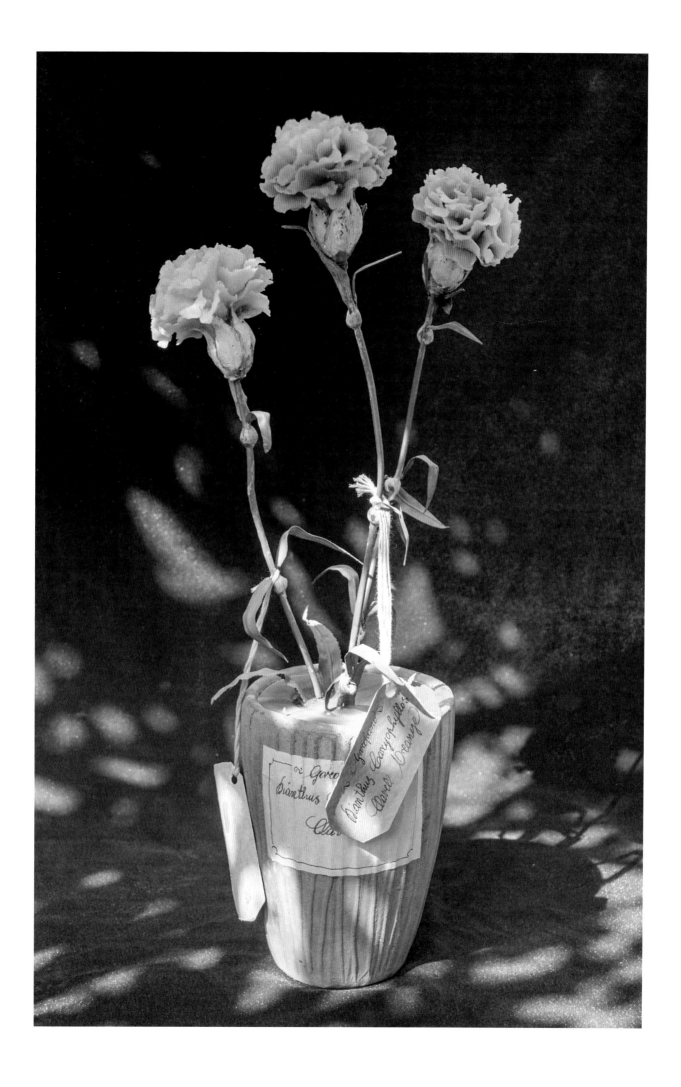

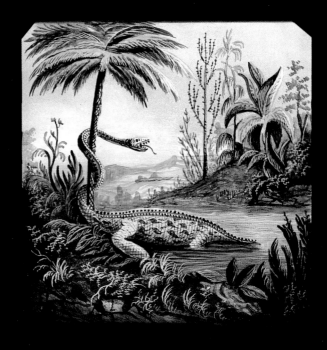

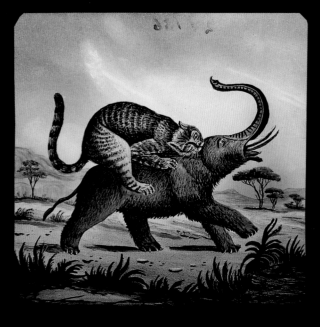

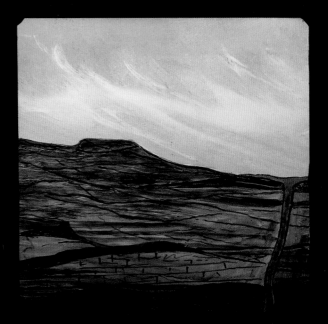

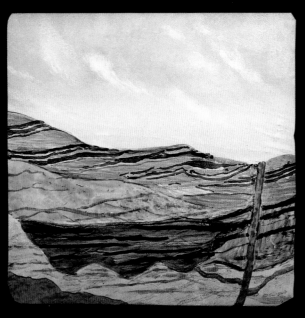

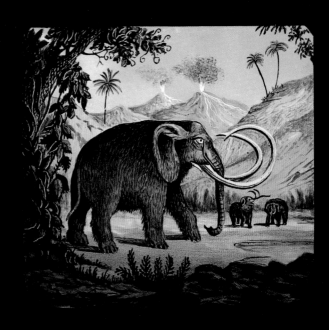

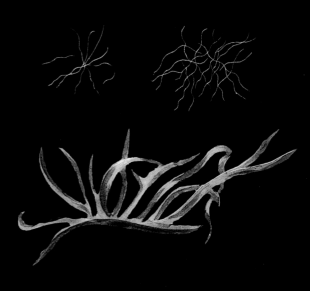

13 magic lantern slides of
extinct animals as well as
geological and natural
history subjects
Mid-19th century
11 frames, each: 18 × 54 cm
(7⅛ × 21¼ in)
2 frames, each: 18 × 38 cm
(7⅛ × 15 in)
France

These magic lantern slides are difficult to date. They deal with palaeontology and early animal forms. Seventeen of the images are of land animals, eleven represent various geological formations and landscapes, four are of fossils and another four of sea creatures.

They are hand-painted on glass and mounted in wooden frames. The glass panels have been secured in the frames with scrap paper, which, I discovered when I had them restored, were scraps of letters in French from the 1830s.

I enquired at various natural history museums but none were able to either date the slides or identify where the images came from. By going through a large number of nineteenth-century books, I eventually identified the work of the French botanist and geologist Pierre Boitard (1789–1859), as the source of these images.

Boitard had been an officer in Napoleon's army during the Hundred Day Campaign but lost interest in the military and became a writer, popularising natural history and writing anonymously and copiously on natural history, botany, palaeontology and gardening. He also wrote articles on art and short stories for magazines.

In 1836–37 he produced, for the magazine *Musée des Familles – Lecture pour tous* (Family Museum – Reading for Everyone), two lengthy articles about palaeontology entitled 'Paris Before Mankind', written in a very popular form as a travelogue into the remote past. His descriptions and illustrations were based on the work of Georges Cuvier in Paris and William Buckland and Gideon Mantell in England.

Stylistically the images could be from 1820–30 but, as Boitard's articles were reprinted in 1861 in a single volume just after his death, it is possible the book rather than the magazine articles were the inspiration for the anonymous maker of these slides. Given their condition when I acquired them, they had been well used over time.

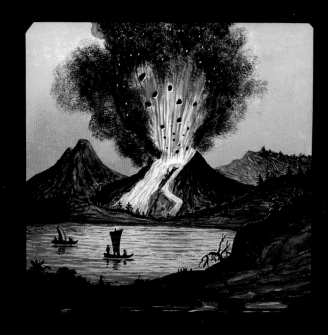

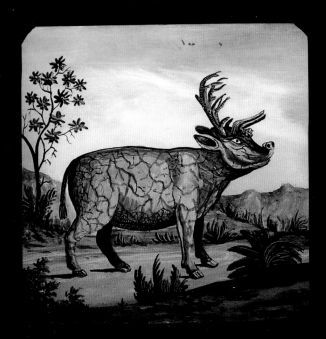

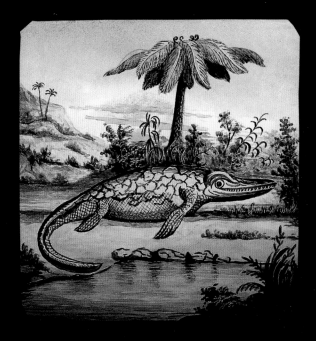

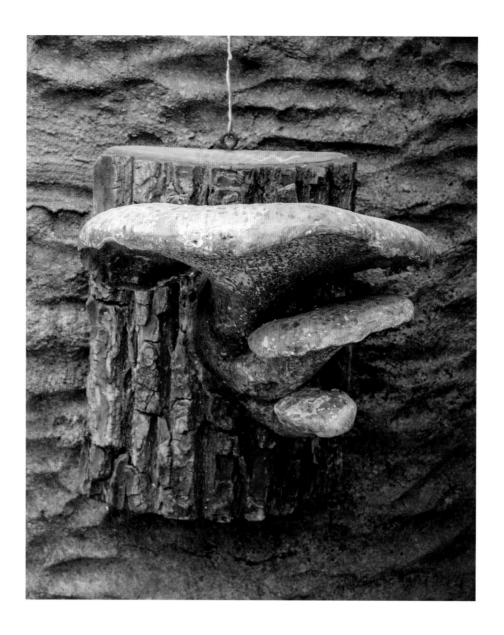

Plaster model of *Polyporus squamosus*
19th century
25 × 18 × 20 cm
(9⅞ × 7⅛ × 7⅞ in)
Germany

This fungus, commonly known as Dryad's Saddle or Pheasant's Back mushroom, is usually attached to dead logs or stumps or clings as a parasite on living trees. It plays an important role in decomposing wood.

This handsome model is clearly from an academic or museum collection. It has a label on the back with the fungus's name and an inventory number in a distinctive late nineteenth-century style.

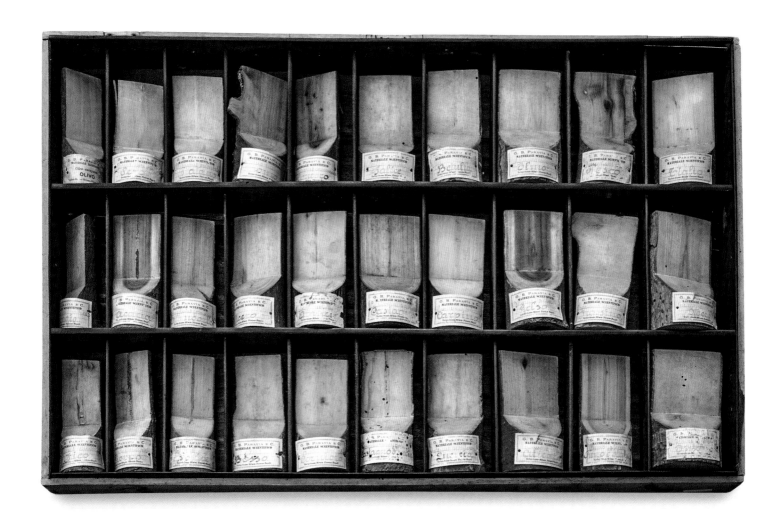

GB Paravia & Co.
30 natural wood samples
Late 19th century
Box: 10 × 72 × 44 cm
(4 × 28⅜ × 17⅜ in)
Turin

There is something very satisfying in having 30 similar specimens showing different woods. There seems to have been something about mid to late nineteenth century Italy that produced many currently desirable items, due possibly to a surge in education after unification. Certainly a number of new agricultural institutes were founded in Northern Italy during this period.

This collection of wood samples – one of a number of Italian botanic items in the collection – was clearly put together for didactic purposes. The maker, GB Paravia & Co., was a Turin-based firm dating back to the seventeenth century, which was involved in the publication and selling of popular and scientific books, as well as teaching materials such as this collection of samples. It appears that the same family is still involved in this business today.

Richard Lawrence
The complete farrier, and British sportsman: containing a systematic enquiry into the structure and animal economy of the horse, the causes, symptoms, and most-approved methods of prevention and cure of all the various diseases to which he is liable: with some select and approved original recipes for various diseases
1816
Height: 27 cm (10⅝ in)
Birmingham

Ten models of horses' jaws
Late 18th or early 19th century
Box: 31 × 42 × 10 cm
(12⅕ × 16½ × 4 in)
England

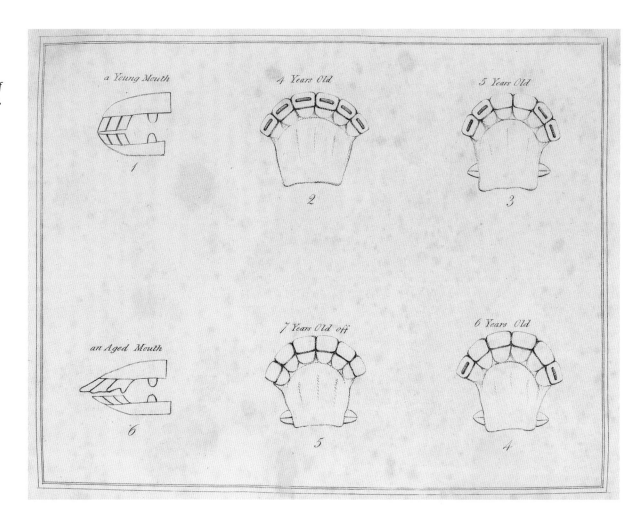

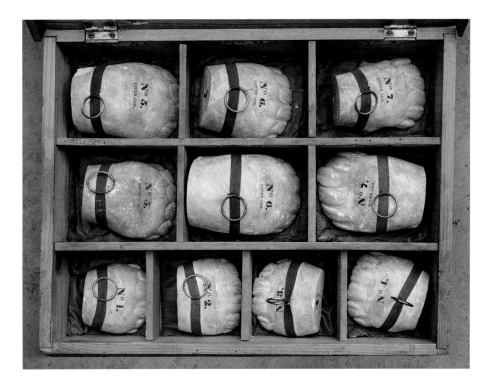

Whereas Richard Lawrence's earlier book *An inquiry into the structure & animal oeconomy of the horse* (p.122), deals exclusively with the horse, this book, published 15 years later, ensures that it is of a wider interest to English gentlemen as it also covers hunting dogs, and details their breeds and characteristics.

The boxed set of models shows the growth of the teeth and jaws of horses ranging from three years old to old age that are also illustrated in the book above. Each model identifies the age of the animal. I can see that these models might have had real practical use when held up against the mouth of a horse. They come from the contents of James Watt's household, sold at auction in 2003.

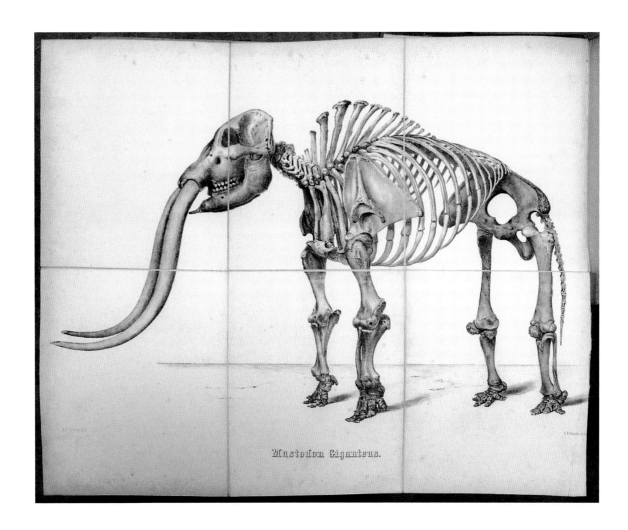

Mastodon Giganteus.

John Collins Warren
*The Mastodon giganteus
of North America*
1855
29 leaves of plates by
Benjamin Franklin Nutting
Height: 34 cm (13⅜ in)
Boston

The bones of a complete skeleton were discovered along the Hudson River in 1845. Other remnants of mastodons had been found in the United States in the eighteenth century.

Originally these animals were thought to be mammoths, similar to those found in Siberia, however Georges Cuvier identified the differences and gave these newly discovered creatures the name Mastodon. Mammoths were members of the elephant family and were grazers with larger, more curved tusks. Mastodons, which were smaller than mammoths, were a distinct species and were browsers of leaves and twigs. The specimen described by John Collins Warren (1778–1856) was displayed in Boston until the beginning of the twentieth century, when his entire collection was purchased by the American Museum of Natural History in New York.

Warren was a very famous Bostonian surgeon, who was Dean of Harvard Medical School and one of the first surgeons to apply anaesthesia.

The illustrations in this publication by Benjamin Franklin Nutting (c.1803–87) are very fine. Nutting was an artist who mainly painted portraits and also produced lithographs. He had frequent exhibitions of his work. The fact that he was more of a professional artist rather than a jobbing illustrator, might explain why the illustrations for this book are so exquisite.

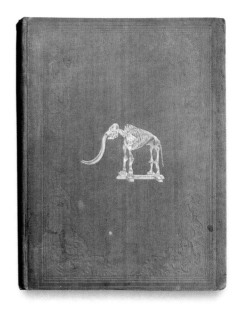

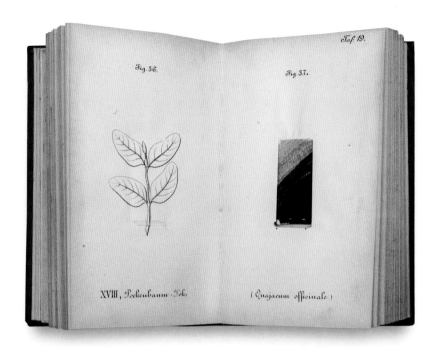

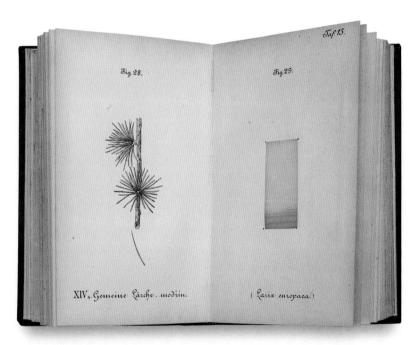

Wenzel Sykyta
Das Holz, dessen Benennungen,
Eigenschaften, Krankheiten und
Fehler: ein Leitfaden zum
leichten Erkennen einzelner
Holzarten und eines
schadhaften Holzes für
Eisenbahn-, Gruben-, Forst-,
Holz-, und Civiltechniker sowie
Bau- und Zimmermeister
1882
168 plates and 25 original
wood samples
Height: 18 cm (7⅛ in)
Prague

Xylotheques are wood collections which take various forms, from straightforward collections of loose or boxed wood samples (see p.213) to assembled and painted panels or boxes (see p.201) or book-like collections like this one and Romeyn Hough's book of American wood species (see p.121).

This small bound book from the Czech Republic deals with various aspects of wood, names of species, their characteristics, diseases and imperfections and their use for various purposes such as industry, railways and construction. It has illustrations of the leaf of each tree described and very finely cut samples of various species of trees inserted into the pages. It principally covers useful wood species.

There are some 80,000 species of tree in the world. Only around 450 species are used commercially and have been properly studied. Some 8,000 tree species are threatened with extinction.

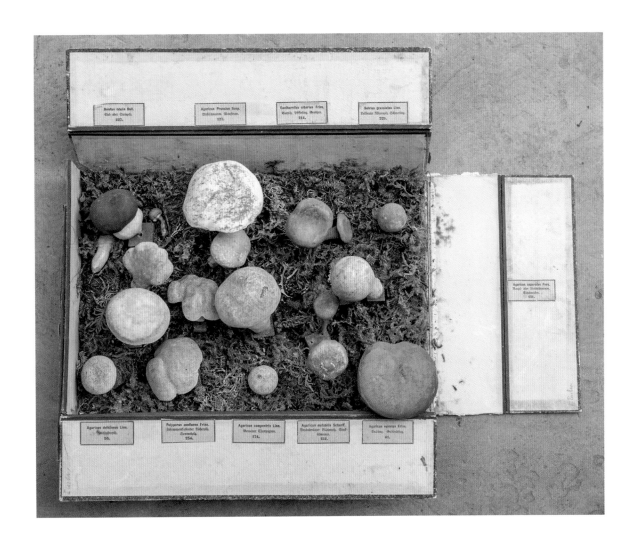

Prof Büchner
18 plaster models
of mushrooms
c.1870
Box: 15 × 42 × 22 cm
(5⅞ × 16½ × 8⅝ in)
Germany

These are painted plaster fungi set in a bed of real
dried moss, with labels in Linnaean and common
nomenclature. This is a classic set, produced for
teaching purposes, which could be safely stored
when not in use.

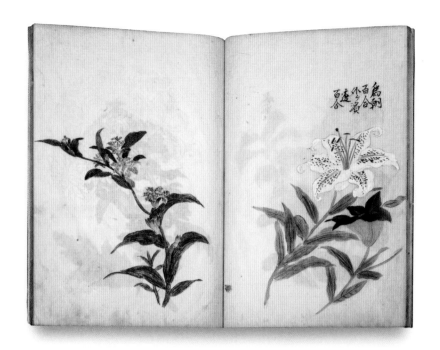

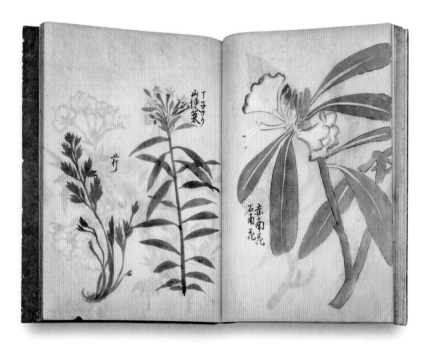

Shasei Noyama-Gusa
1832
2 vols. in portfolio
Height: 22 cm (8⅝ in)
Japan

This set of two books (the title of which can be roughly translated as 'Sketchbooks of Wild Flowers') dating from 1832 is quite an early item in relation to the rest of the collection. The illustrations are very delicate and other than the odd plant name there is no other text to provide information, so this is more of an aesthetic than a scientific publication. Quite soon after the publication date of these sketchbooks botanical illustrations in Japan became more scientific. These watercolours are on the edge between the two: a moment that often provides remarkable work.

Display of pufferfish and
marine specimens
19th century
40 × 43 × 29 cm
(15 3/4 × 16 7/8 × 11 3/8 in)
England

The pufferfish is part of the *tetraodontiformes* family,
a name that refers to the four large teeth fused into
an upper and lower plate, which these creatures
all have for crushing crustaceans and molluscs.
The pufferfish is generally believed to be the second
most poisonous vertebrate after the golden poison
frog. Certain internal organs such as its liver are
highly toxic and their preparation in Japan and
China requires skilled chefs. Its name derives from
the fact that it can also blow itself up as a method
of defence.

In general, I am not particularly interested in
taxidermy because it often serves more as
decoration than as a didactic tool, but occasionally,
as seen in this case where it is displayed in its
handsomely arranged Victorian box, I am charmed
by a particular specimen.

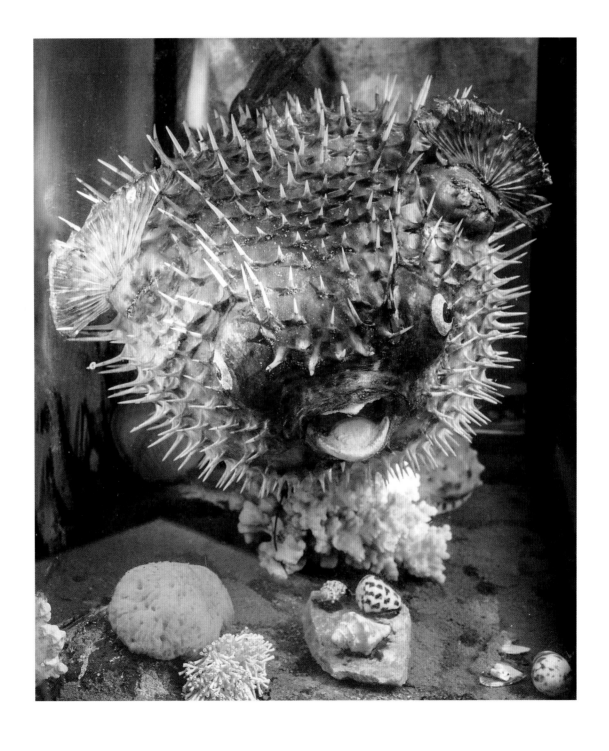

Francesco Garnier Valletti
Two boxes of wax fruit
19th century
Top: 10 × 51 × 41 cm
(4 × 20⅛ × 16⅛ in)
Bottom: 10 × 40 × 37 cm
(4 × 15¾ × 14⅝ in)
Turin

These boxes of lemons and peaches are remarkable objects. They demonstrate diseased fruits of different varieties and were clearly made to teach students how to avoid imperfections in horticulture. These come from an agricultural college in Northern Italy, as do the wax models of carnations (pp.208–09) and pomegranates (p.237).

These models were most likely made by Francesco Garnier-Valletti (1808–89) who lived principally in Turin and made many collections, including his own, which can still be seen there in the Museo della Frutta.

The labels on the fruit are more recent, as are the boxes. These are of aluminium and were made some time in the early twentieth century. All this leads me to suppose that these models were handled regularly, possibly well into the twentieth century, rather than being put on display, or – even worse – into storage, and so their didactic function was lost quite late. Maybe Italians are more conservative, or – more likely – there was no money to adopt more modern techniques.

Today we can look at these fruits removed from their original context and attribute to them any virtues we choose.

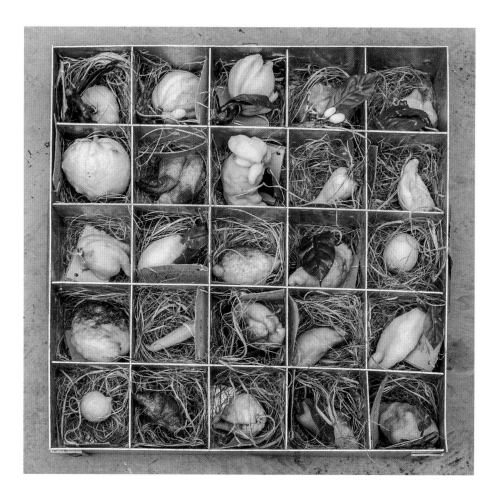

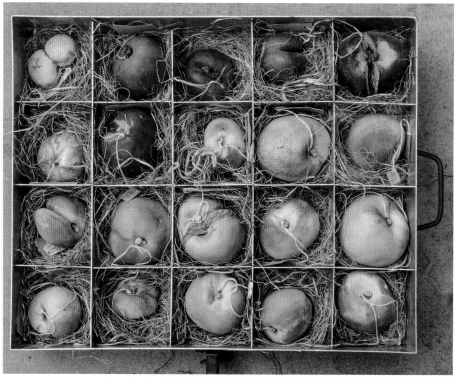

James Deane
Ichnographs from the sandstone of Connecticut river
On stone from nature by J Deane
1861
46 leaves of plates
Height: 31 cm (12¼ in)
Boston

James Deane (1801–58) was not an academic. He was the son of a farmer and decided at some stage to study medicine, although he did not go on to practice very successfully as a doctor.

Deane took an interest in the footprints he found in sandstone slabs used as paving stones in his hometown in Connecticut and attributed these, as others did, to birds. This book was eventually published posthumously. It features lithographs and photographs stuck onto the pages, and also includes plant impressions, insect tracks and raindrops as well as footprints. It would appear that Deane was more interested in the fossils' wonderful shapes and forms than in the significance they held in the field of palaeontology.

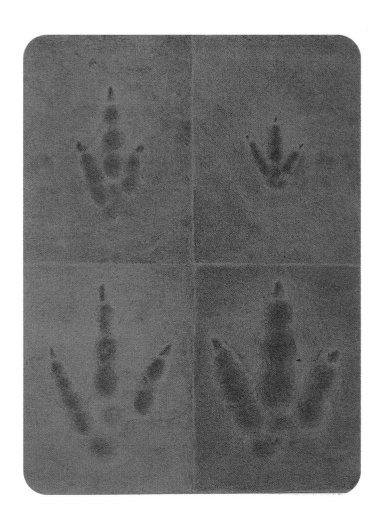
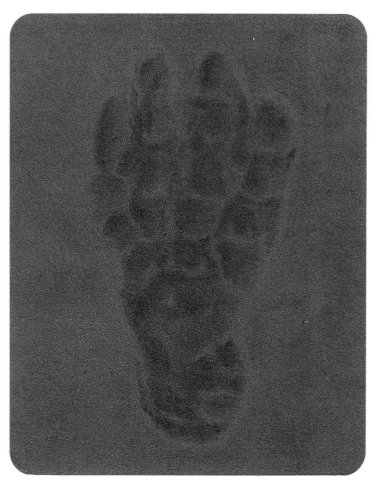

Johann Gottlob von Kurr
The mineral kingdom;
with coloured illustrations
of the most important minerals,
rocks, and petrifactions
1859
24 leaves of plates
Height: 36 cm (14⅛ in)
Edinburgh

Johann Gottlob von Kurr (1798–1870) was a prominent professor in Stuttgart. This work was a translation from the German version that he had published a year earlier.

Both the German and the English editions are particularly finely illustrated. The illustrations are hand-coloured copper plates, some heightened with gum arabic, while for others metallic ink has been used on the plates themselves. Unfortunately the name of the artist is not known.

The dealer from whom I bought it wrote on the inside cover: 'some of the best nineteenth-century hand colouring to be found anywhere. A rare work regularly broken for the beauty of its plates'. That just about tells you everything about what happens to these books.

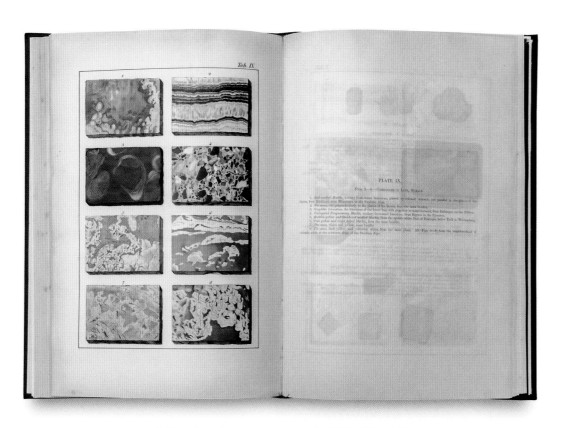

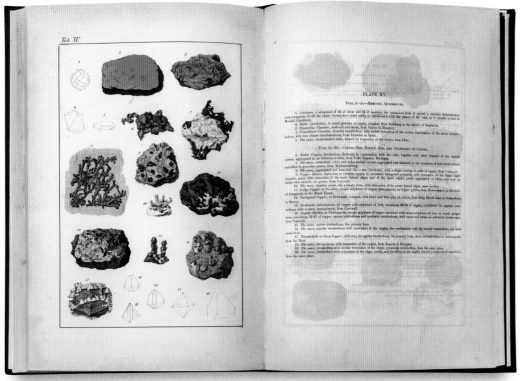

Album of butterfly wings
1862–67
37 × 32 × 4 cm
(14⅝ × 12⅝ × 1⅝ in)
Mexico

These impressions used nature-printing, where the wings of the butterflies were treated with a liquid preparation to release the colours onto paper. The body, antennae and legs were then painted in. This unusual technique resulted in remarkable colours and textures.

The individuals involved in the production of this work are not recorded but the album was part of a scientific expedition organised by Napoleon III (1808–73) during the French adventure in Mexico from 1861–67.

Napoleon III took an interest in Mexico, hoping to benefit from the country's resources and taking the opportunity of chronic anarchy, the Mexican suspension of the repayment of international debt and probably also from the United States' distrac-

tion with its own civil wars, which lasted from 1861 to 1865. Napoleon put the Austrian Habsburg Ferdinand Maximilian (1832–67) on the throne in 1864, but then abandoned him in 1867. The artist Édouard Manet (1832–83) famously painted Maximilian's execution by the Mexicans. His wife Charlotte (Carlota) (1840–1927), daughter of Leopold I of Belgium, returned home to seek support for her beleaguered husband, but went mad and remained so until she died in 1927.

The scientific expedition to Mexico covered fields including natural history, medicine, public health, physics and chemistry, history, linguistics, archaeology, economics, statistics and public works. The findings of these investigations were published in stages up until 1899.

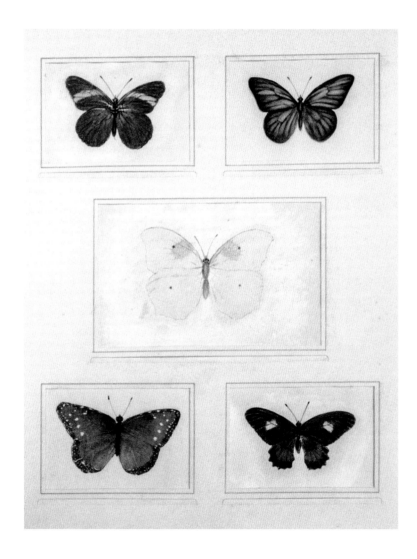

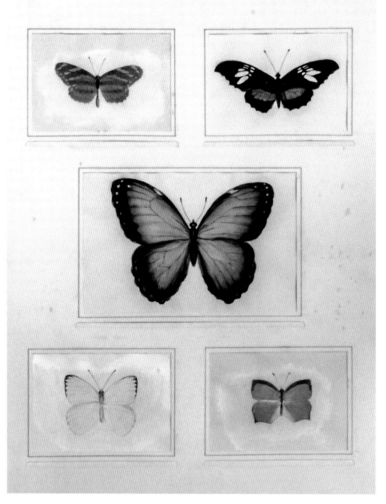

Eric Craig
Boxed collection of
fern specimens
Mid-19th century
Box: 14 × 27 × 19 cm
(5½ × 10⅝ × 7½ in)
New Zealand

In New Zealand ferns were not only plentiful but
have became emblematic. The 'All Blacks', the
national rugby team of New Zealand, display a
fern on their rugby shirts.

This box is beautifully made, and features 125
individual specimens mounted on 22 × 14 cm cards
(8⅝ x 5½ in). Its creator was Eric Craig (1829–1923),
a well-known collector, publisher and dealer of
natural history and ethnography in Auckland.

Although the ferns are identified with formal
Linnaean names, this box was probably intended
more for the amateurs who were part of the
Victorian craze for ferns, than for scientists. But
the amateurs clearly wanted the real thing.

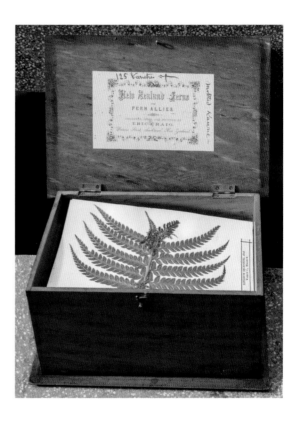

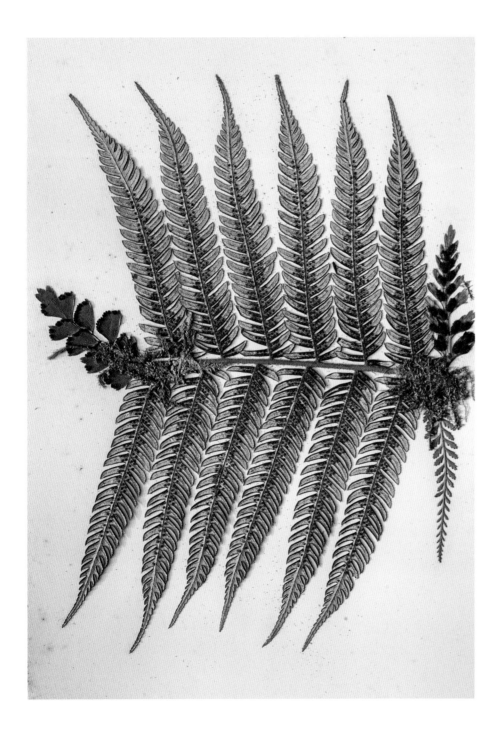

Henry Smith
Specimens of nature printing
from unprepared plants, etc, etc.
1857
Height: 54 cm (21¼ in)
Madras

This is one of the great nature-printing books,
produced in India during the days of the British East
India Company, which uses the original technique
of coating a flat object with ink and printing the
object onto paper. This book is fairly unique in that
both sides of the plants, and a few snakeskins, are
printed. It was produced in the year the Indian
Mutiny broke out. Henry Smith was superintendent
of the government press in Madras at the time.

Only seven other copies of this book are believed
to be in existence: four in the United Kingdom and
three in the United States.

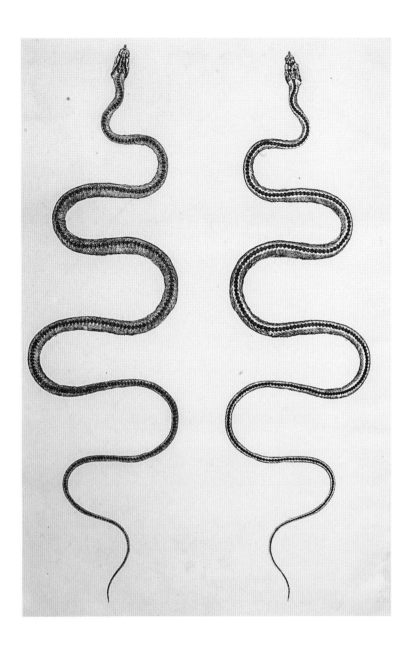

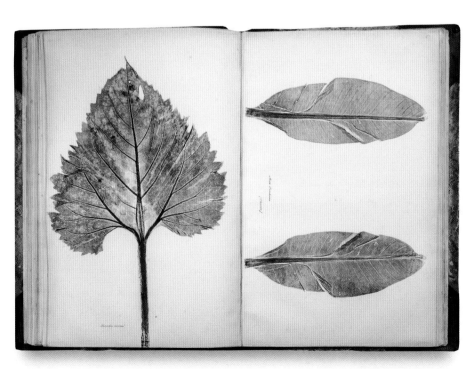

Eight photomicrographs
of fleas
Late 19th century
Each: 18 × 13 cm (7⅛ × 5⅛ in)
France

It is not clear who made these enlarged images
of fleas, nor when exactly they were photographed
but they have a certain charm. Their history and
associated patina – including labels, handwritten
numbers, evidence of handling and fingermarks –
convey to me a feeling of nostalgia. They are
labelled with two sets of index numbers, probably
from the early part of the twentieth century.

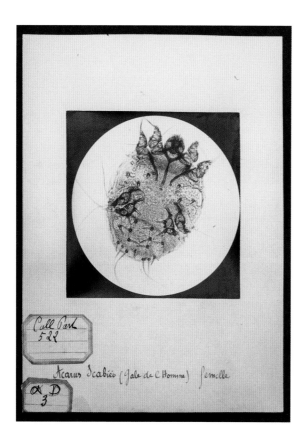

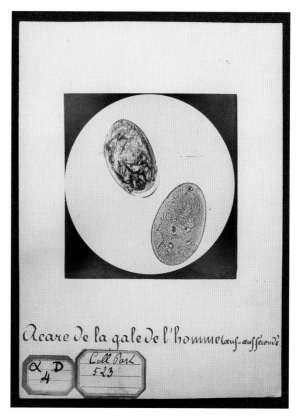

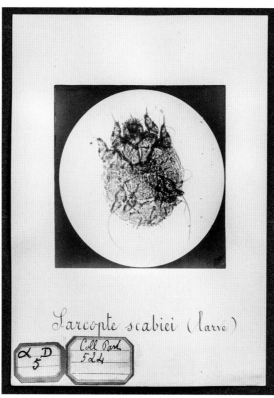

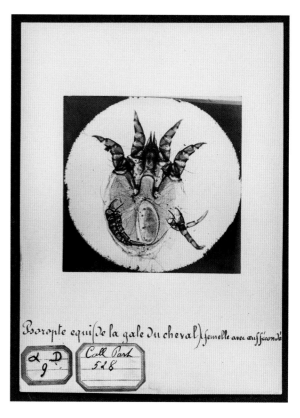

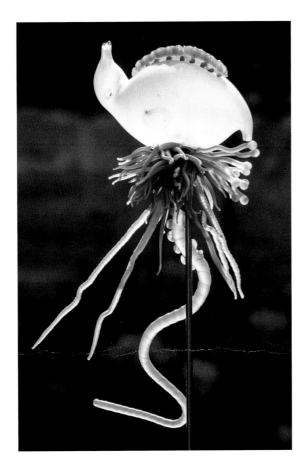

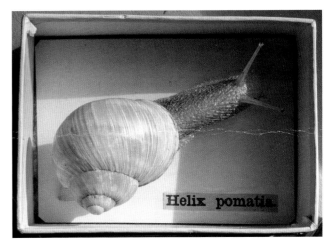

Helix pomatia.

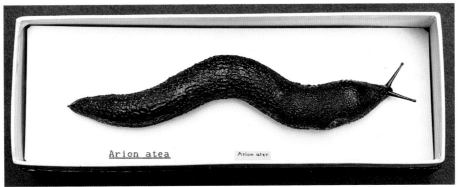

Arion atea   Arion ater

Leopold and Rudolf
Blaschka

Glass model of a Portuguese
man o'war
Mid-to late 19th century
14 × 8 × 8 cm
(5½ × 3⅛ × 3⅛ in)
Germany

Ten glass models of slugs
and snails
Mid-to late 19th century
Slugs
Each, between:
6–15 × 1–4 cm (2–6 × 1–2 in)
Snails
Each, between:
2–9 × 1–4 cm (1–3 × 1–2 in)
Germany

I exhausted the London dealer's stock by buying up his last Blaschkas: ten models of slugs and snails and a wonderful Portuguese man o'war. Although I was not aware of anyone else trying to buy these models, the negotiations were quite protracted. It is not easy to put a price on objects when there is no history of a market.

It is worth noting that when making the snails the Blaschkas were happy to use the real shells and make glass bodies for them. They are highly realistic and did the job, which was to demonstrate the species.

This Portuguese man o'war is, of course, much smaller than in real life where the long thin tendrils can extend to 50 metres in length below the surface.

Although it is often mistakenly considered a jellyfish, the venomous Portuguese man o'war is in fact a siphonophore, which means it is a group of tiny zooids making up a colony of organisms that can only survive by working together. Its name derives from the appearance of its upper-most polyp, a gas-filled bladder, which sits above the water – and resembles an eighteenth-century Portuguese fighting ship in full sail. The tentacles are covered with venom-filled organisms used to paralyse and kill fish and other small creatures.

This little model is one of the treasures of the collection. It is like a piece of jewellery, which is why it is included in the first part of the book (p.59).

Album of seaweeds
19th century
24 × 20 × 5 cm
(9½ × 7⅞ × 2 in)
England

I rescued this little bound collection of dried British
seaweeds from a dealer who was about to cut it
up and frame the individual pages as decorations.
The seaweeds inside are very beautiful but what
struck me was the *ex libris* plate, which stated that
the collection had belonged to a Sir Oswald Mosley,
Baronet. In a way this looked too good to be true.
Did the famous British fascist Sir Oswald Mosley,
6th Baronet (1896–1980) really own this little book?
It is more likely in fact that it belonged to one of
his ancestors, of whom there were three named
Sir Oswald. The first was knighted in 1720, the
second died in 1757 and the third in 1845.

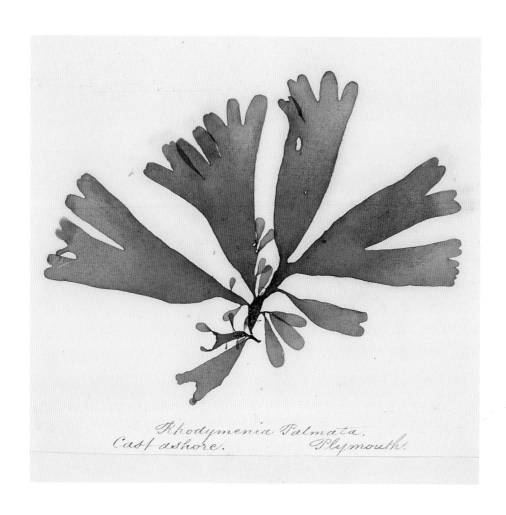

Fred Enock
Diagrams of insects
Late 19th century
Each: 60 × 50 cm
(23⅝ × 19¾ in) (approx.)
England

Fred Enock (1845–1916) was a Quaker and draughts-man for a civil engineering firm who worked on Blackfriars Bridge in London. His hobby was natural history and he eventually went to work for his uncle who supplied microscope slides. In 1876 they had an inventory of 20,000 slides. Enock became an expert at making insect slides and began to give lectures on entomology, projecting his own microscope slides and photographs from nature.

These 24 drawings – accompanied here by two texts – were made to be photographed and converted to lantern slides for lectures. Like most teaching diagrams, they emphasise the schematics of what they are demonstrating, so one often ends up with something quite unusual.

These drawings are all about earwigs. Who would have thought that their exaggerated features would make them so beautiful?

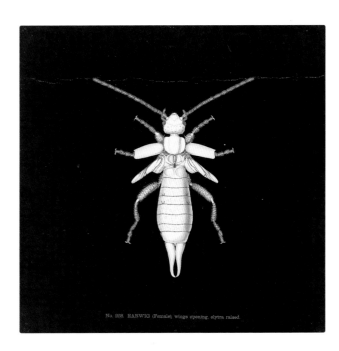

No. 238. EARWIG (Female), wings opening, elytra raised.

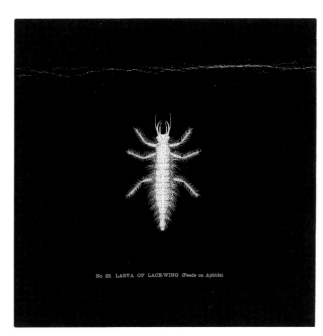

No. 22. LARVA OF LACE-WING (Feeds on Aphids).

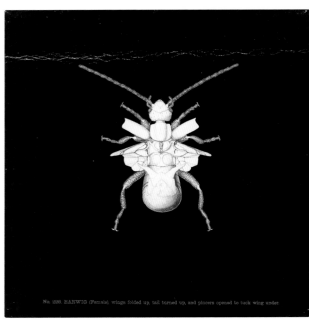

No. 226. EARWIG (Female), wings folded up, tail turned up, and pincers opened to tuck wing under.

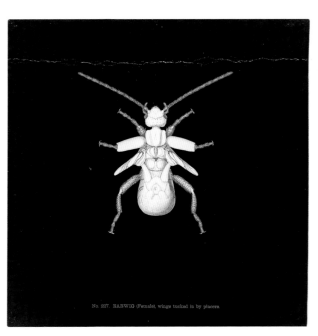

No. 227. EARWIG (Female), wings tucked in by pincers.

Two mineral spar towers
19th century
91 × 41 cm (35⅞ × 16⅛ in)
Not illustrated:
45 × 31 cm (17¾ × 12¼ in)
Europe

These two spar towers both come from an auction in London in 2012 of the contents of a private house in Italy. The one illustrated is, at nearly one metre, quite unusually tall. By the nineteenth century, natural history objects like these had become part of fashionable decoration. I suspect the original owner was still interested in the mineralogical aspects of these spar towers.

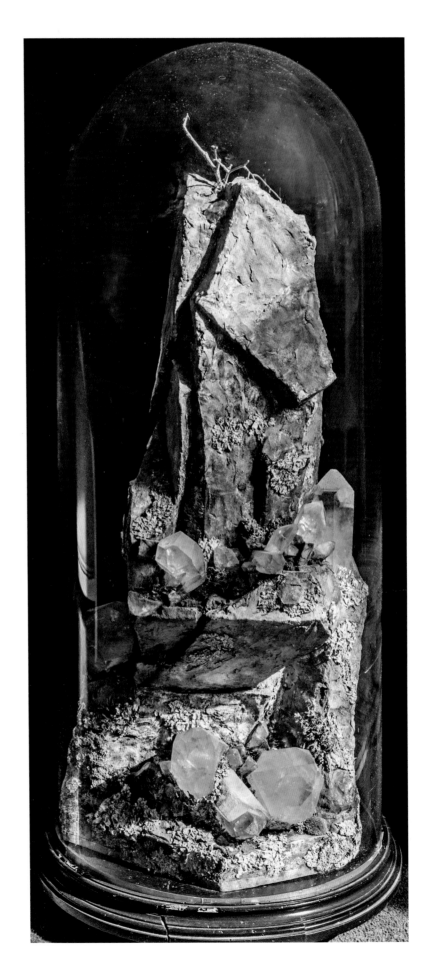

Six papier-mache models
of mushrooms
Mid-to late 19th century
Each, between:
9–24 × 12 × 12 cm
(3½–9½ × 4¾ × 4¾ in) and
24 × 12 × 12 cm
(9½ × 4¾ × 4¾ in)
Germany

These mushroom models are typical teaching models with labels in Latin and Czech. They represent the varieties *Lactarius volemus*, or weeping milk cap; *Leccinum rufum*, the red-capped scaber stalk; *Candarellus vulgaris* (*Cantharellus cibarius*) or chanterelle; *Ptychovera bohemica* (*Verpa bohemica*), the wrinkled thimble-cap; *Craterellus Nigrescens* (*Craterellus cornucopioides*) or the horn of plenty mushroom; and *Amanita porphyria* or Grey Veiled Amanita.

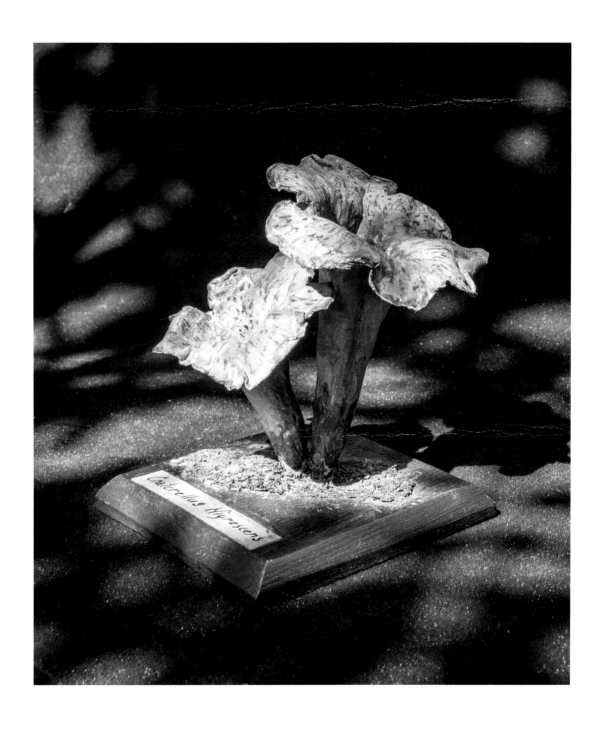

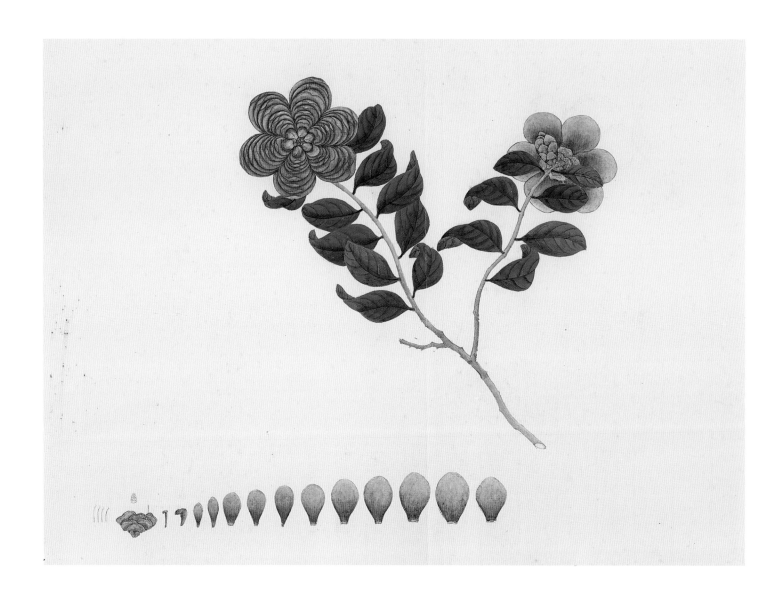

Chinese School
Six botanical studies
Late 18th to early
19th century
4 studies, each: 43 × 59 cm
(16⅞ × 23¼ in)
2 studies, each: 42 × 52 cm
(16½ × 20½ in)
China

These six studies of Chinese flowers in pencil, watercolour and gouache, with pen and ink inscriptions of the flower name in Chinese are typical of the early nineteenth-century plant studies that were produced for European patrons. Four are stamped on the back with 'Haw Sham / Flower Painter' in Chinese characters. As was typical of the time, these paintings by Chinese artists are on English paper: four are on laid or ribbed paper with the watermark 'Stacey Wise' and dated 1817; and the remaining two are on wove or untextured paper with the watermark 'I Taylor' and dated 1794.

These are early examples of what became an enormous interest in Chinese botany in Europe, and particularly in the UK.

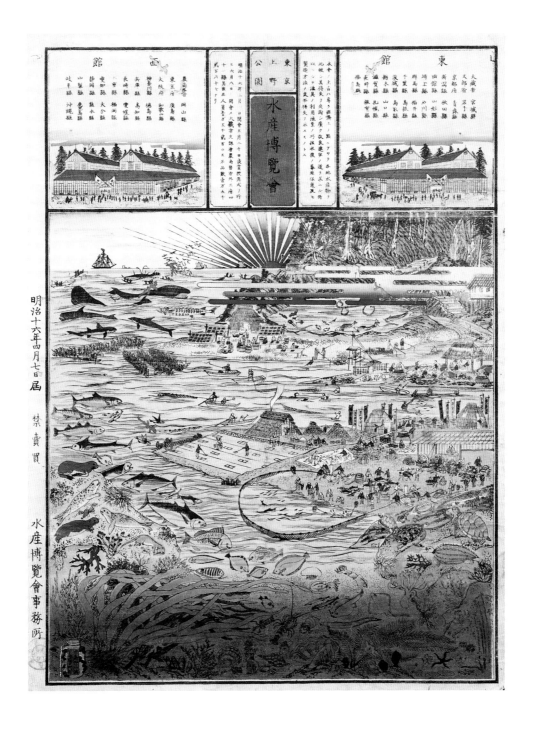

Poster for Aquatic
Exposition
1883
40 × 30 cm (15¾ × 11¾ in)
Tokyo

After the Meiji restoration in 1868 all sorts of
attempts were made to modernise commerce and
industry in Japan. The poster illustrated refers to
an exhibition organised in Ueno Park in Tokyo
from 1 March to 8 June 1883, which was the first to
promote fishery activities as well as stimulate
demand for regional products.

The text along the top of the poster describes
which government and regional institutions
exhibited in the East and West Halls.

The exhibition, which is said to have attracted
230,000 visitors, including the Emperor, is reputed
to have been one of the projects of Yoshio Tanaka
(see p.95). Tanaka had served under the pre-Meiji
regime, then spent time in Paris and became a
successful civil servant and a promoter of practical
natural sciences.

Carved wooden face-
reading model
19th century
32 × 20 × 17 cm
(12⅝ × 7⅞ × 6¾ in)
China

After the communist takeover in China in 1949, material related to this type of activity – face-reading, which was a method of fortune telling – was frowned upon as part of the 'old orders: old mindsets, old habits, old customs and old culture'. As a consequence, items like this head are no longer to be found in China. This particular example came from a dealer in London.

My friend Ying Sheng Yang, an artist from mainland China living in London, has some interesting things to say about this head. Ying tells how in 1986 he went to say goodbye to his grandmother in remote northern China before he left to come to England. He and other passengers on a small train were stranded at a rural railway station when they came across an old-fashioned fortune-teller, who looked each person in the face and accurately revealed key aspects of his or her life.

The writing on the face of this model consists of combinations of two words that together form a sort of code dealing with aspects of an individual's life. The practitioner has to know and interpret the code. The model identifies where to look on the face for various aspects of an individual's life, past and future. For example, the area below the left eye, from left to right, refers to the 'eldest', 'middle' and 'younger' son while the area below that refers to brothers and nephews. Under the right eye is a series of numbers, of which the meaning is unclear.

Two important points are marked: the centre of the upper lip is believed to be the middle point of a human being and, for this reason, people in China push hard on this point if someone has fainted, as a kind of starter button. Another is the centre of the forehead, the place of the soul – which is known as 'heaven's gate' – and which on the model shows wear and tear. The chin refers to the ancestral home and to other domestic arrangements, such as servants. Above the right ear lies an area dealing with desires while the area above the left ear deals with 'low level principles', which could be interpreted as standards that are not too difficult to achieve. This area is also somewhat worn.

I chose to include a detail of part of this head in the first part of the book (p.9) because of references to their history that I see in the work of contemporary Chinese artists.

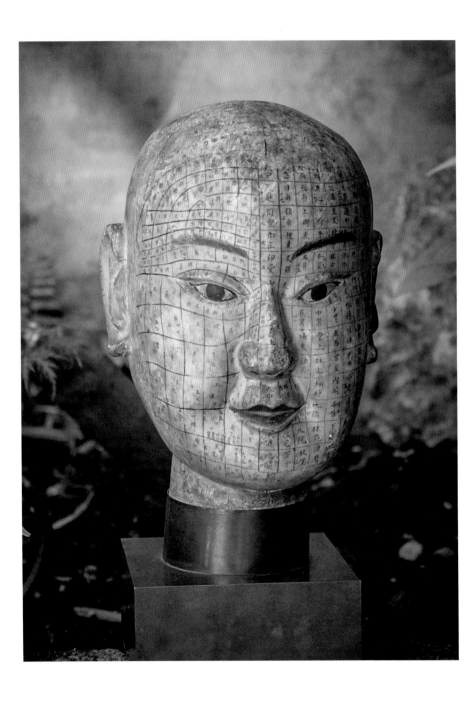

Louis Agassiz
Recherches sur les poissons fossiles
1833–45
436 colour lithographs
by J Dinkel, Nicolet,
Jeanjaquet and A Sonrel
12 vols.
6 vols. of text
Each, height: 34 cm (13⅜ in)
6 vols. of plates
Each, height: 29 cm (11⅜ in)
Neuchatel

Louis Agassiz (1807–73) was a Swiss naturalist who ended up at Harvard University where he founded the Museum of Comparative Zoology. When he was still a young man working in Paris, the French naturalist and zoologist Georges Cuvier asked him to work on fossilised fish. This work was published between 1833–44 in 12 volumes. He employed one of my favourite artists, Joseph Dinkel, who also worked with Richard Owen (p.123), to make most of the illustrations.

Agassiz sold the original watercolours in England when he ran out of money. They are now in the Geological Society in London. The originals, which are in most cases quite large, are even better than the lithographs in Agassiz's book (see p.57).

Agassiz today has a mixed reputation. He was anti-Darwinian and appeared to choose the wrong side in racial arguments at the time of the US Civil War.

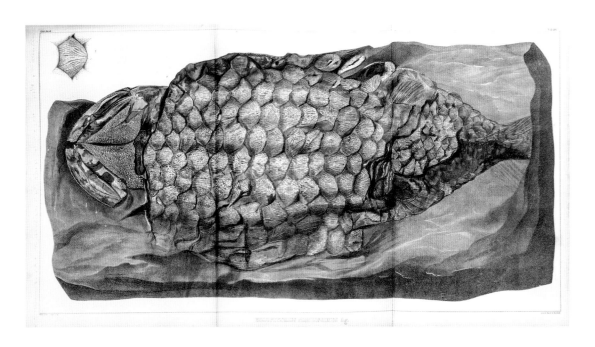

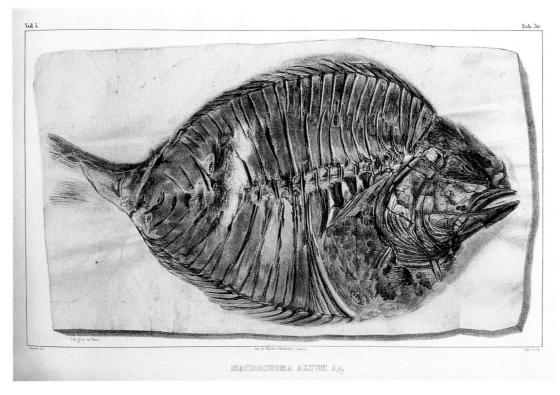

12 boxed wax fruits
Early 20th century
8 boxes, each: 5.5 × 9 × 9 cm
(2⅛ × 3½ × 3½ in)
4 boxes, each: 5.5 × 17 × 9 cm
(2⅛ × 6¾ × 3½ in)
Belgian Congo (Democratic
Republic of Congo)

These fruits in boxes reputedly come from the Belgian Congo, which would date them to some point in the 1920s or 1930s. On close inspection they are handsomely made. It is a pity that so little is known about their provenance, but as objects, they hold a definite appeal.

This set contains a combination of species that are native and non-native to this area of Africa. The only confirmed native species is *Ricinodendron heudelotii* (Njangsa). Also included are four species from the genus *Annona*, which are unlikely to be native and were probably imported from the Americas. The remaining ten fruits were certainly imported: guava, cherimoya, custard apple, malabar chestnut, sweet pepper, hog plum, banana, water apple, mango and orange. All these imported fruits are now cultivated in West Africa.

The fact that the wax has been exposed to what looks like heat suggests that these models spent time in the tropics.

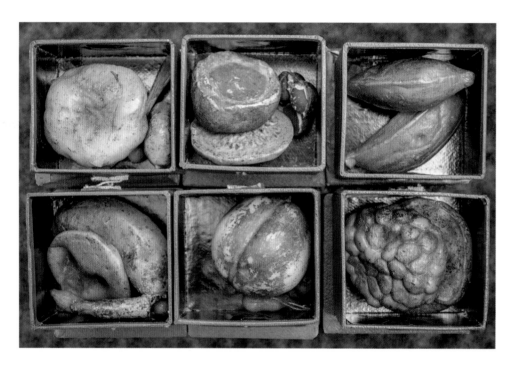

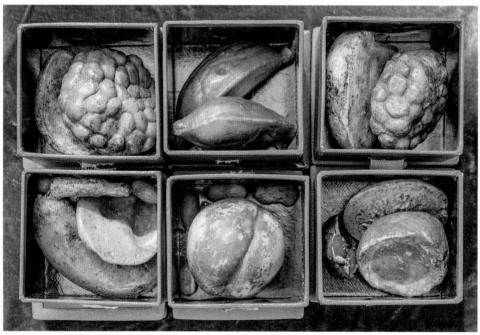

Francesco Garnier Valletti
Two wax models
of pomegranates
19th century
Each: 8 × 9 × 11 cm
(3⅛ × 3½ × 4⅜ in)
Turin

These pomegranates come from the same Northern Italian agricultural college as the wax carnations and the deformed lemons and peaches (pp.208–09, 220), and also show diseased or imperfect fruit. They have no labels. The model maker has captured the very fleshy insides of the pomegranates. I asked Rosamond Purcell to photograph them like gaping wounds (pp.52–53). Goya lives on…

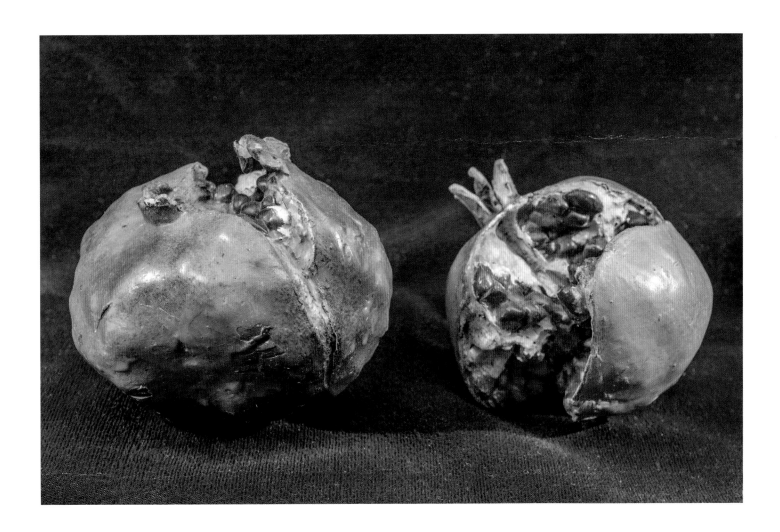

The plate illustrated is the result of the two-stage nature-printing plate-making process that was perfected by the Austrians, in the mid-nineteenth century, in order to produce multiple copies from the same specimen (see p.107).

In the early days of nature-printing printers produced single copies of a specimen by coating it with ink and then using this to print with. To produce more than one copy they required a durable printing plate and for this a soft impression into lead or gutta percha, a rubber-like substance, had to be converted into a hard plate that could be printed over and over again.

The soft impression was coated with graphite and immersed in a liquid solution of copper sulphate through which a low-voltage electrical current was sent, depositing the copper onto the lead impression. However, in order to print, the impression had to be in the plate in order to hold the ink. The intermediate plate shown here has the impression raised above the flat copper, not etched into it, so the process is repeated and a final plate is made.

It is quite unusual to come across one of these intermediate plates. I found this example in Vienna, where this process was invented.

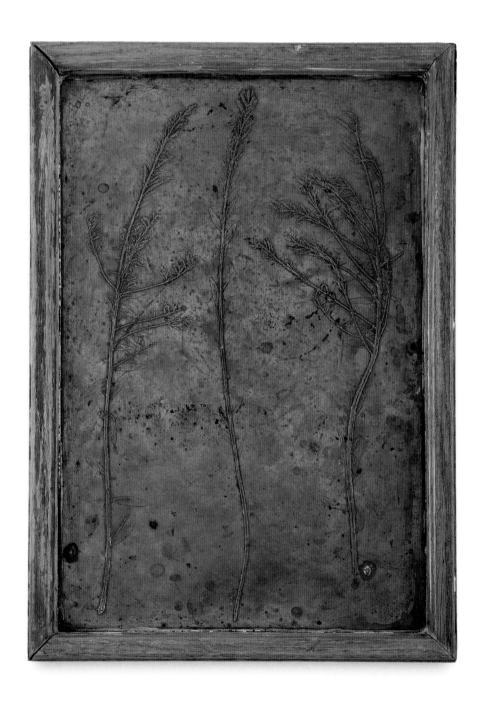

John Milne & Sons
The British farmer's plant portfolio: Specimens of the principal British grasses, forage plants and weeds: with full description
1890
56 leaves of mounted dried plants
Height: 23 cm (9⅛ in)
Montrose

John Milne & Sons, seedsmen and seedgrowers from Montrose in Scotland, produced this portfolio of dried grasses to help identify the best forage plants for producing animal feed. In the second half of the nineteenth century Britain was going through a significant conversion of farming land, from crop production to stock-raising, as a consequence of the dramatic increase in grain imports from Canada, the United States, Argentina and Russia. Most farmers left the land to go to grass, with no idea of the various species and their worth. At the time a specialist said 'not one of the pastures I have visited is so good as it easily might be made'.

Fifty-six varieties of grass are described and herbarium specimens of each species are handsomely displayed.

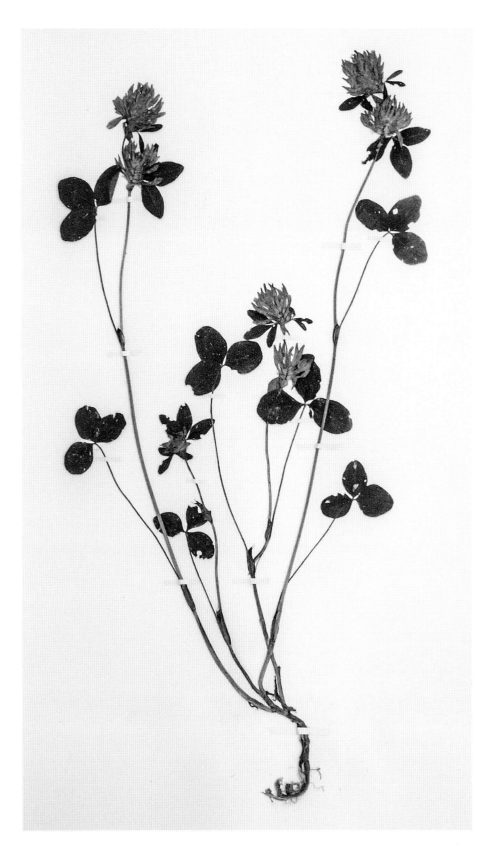

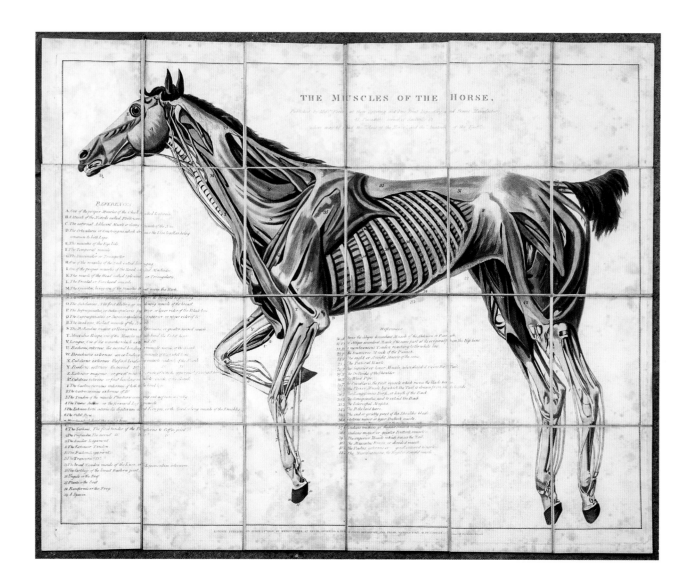

The Muscles of the Horse
1848
Etching mounted on
foldable canvas
25 × 20 cm (9⅞ × 7⅞ in)
(folded)
London

This is another example of a folding illustration that allows a large print to be reduced to a manageable size. It is likely to have been made for sale to vets and sportsmen.

Nothing is known about the artist but the overall effect is quite pleasing.

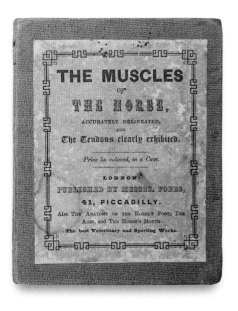

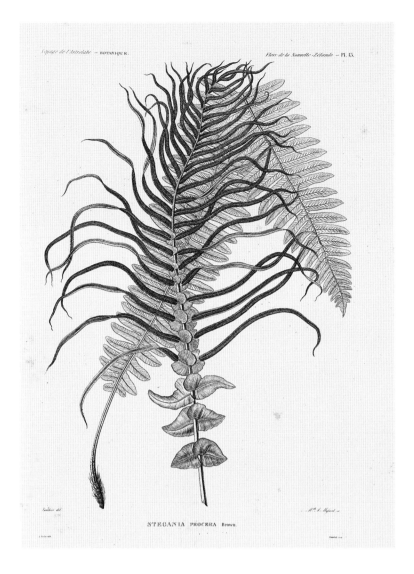

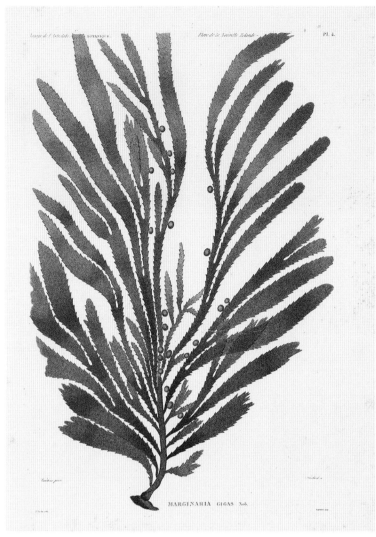

Jules Dumont d'Urville
Voyage de la corvette L'Astrolabe
éxécuté pendant les années
1826–1827–1828–1829 sous
le commandement de M. Jules
Dumont d'Urville, Capitaine
de Vaisseau. Atlas
1833
80 engraved botanical
plates by M Vauthier,
E Delile and others
Height: 51 cm (21⅛ in)
Paris

Having lost most of their colonies in the western hemisphere in the wars with Britain in the eighteenth century, the French set out to explore the Pacific. One forgets how many voyages were undertaken: Bougainville (1766–69), de Surville (1769–70), Marion de Fresne and Crozet (1771–73), de la Pérouse (1785–88), Marchand (1790–92), de Bruni d'Entrecasteaux (1791–93), Baudin and Hamelin (1800–04), de Saulces Freycinet (1817–20), Dupperey and Dumont d'Urville (1822–25), Dumont d'Urville (1826–29), Laplace (1829–32), Dumont d'Urville and Jacquinot (1837–40) and de Rocquemaurel (1851–54).

This botanical atlas comes from a publication describing the first voyage of the French ship *Astrolabe*, which was under the command of the explorer Jules Dumont d'Urville (1790–1842) who had previously travelled to the Pacific on *La Coquille*. The atlas consists of 80 plates: 41 illustrating the flora of New Zealand and a further 39 from his earlier expedition.

The complete report of the *Astrolabe*'s voyage covers a wide range of subjects ranging from geography to geology, botany, zoology, entomology and hydrography (the study of bodies of water).

The *Astrolabe* sailed to Australia, spent considerable time in New Zealand, then went north to Tonga, Fiji and New Guinea before returning to Australia via the Dutch East Indies, which is now Indonesia. It touched on Australia once more and then returned westward to France.

The very fine illustrations are mainly by Michel Vauthier and Eulalie Delile. Delile (1800–40) was a comparatively well-known female botanical artist with a naturalist brother, Alire (1778–1850) who had the distinction of accompanying Napoleon on his expedition to Egypt in 1798. She illustrated Karl Sigismund Kunth's *Revision de Graminées* (1829), which was a work based on plants collected by Alexander von Humboldt (1769–1859) and Aimé Bonpland (1773–1858) between 1799 and 1804 in Latin America.

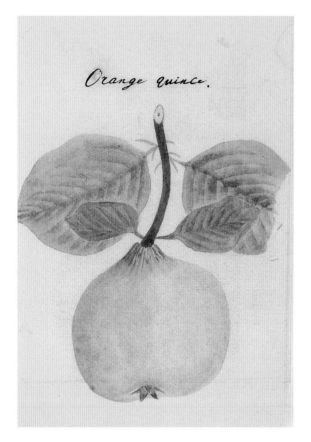
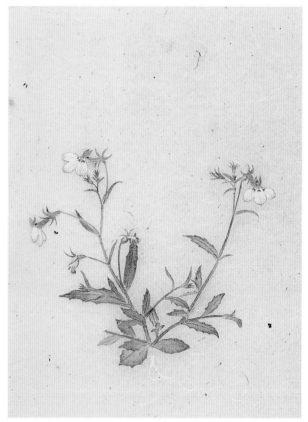

11 botanical watercolours
Each: 25 × 18 cm (9⅞ × 7⅛ in)
Japan

These 11 watercolours are bound in a simple fashion with two pieces of string. They are copies by an unknown Japanese artist of French plants from the *Phytanthoza Iconographia* (1737–45), an important eighteenth century book on plants with more than a thousand engravings, which was compiled by the apothecary and botanist Johann Wilhelm Weinmann (1683–1741).

The copies were commissioned by Yoshio Tanaka, who had been in Paris in the 1860s (see p.95) and would probably have seen Weinmann's famous plant book there. It is not clear if the Japanese illustrator was with Tanaka in Paris or if Tanaka had access to a copy of the book in Japan.

For some reason one watercolour is labelled with English text – 'Orange Quince' – while the others are in Japanese.

Gustave Joseph Witkowski
*Anatomie iconoclastique:
atlas complémentaire de tous
les ouvrages traitment de
l'anatomie et de la physiologie
humaines: composée de
planches decoupées, colorées
et superposées (texte inclus):
Organes génitaux et perinée
de la femme*
1876
Height: 37 cm (14 ⅝ in)
Paris

Gustave Joseph Witkowski (1844–1922) was a French surgeon known for his books on medicine and anatomy.

This little publication is another 'flap book', where the body can be 'dissected' by lifting each layer of flaps. The process is described here as '*planches découpées, colorées et superposées*' (cut out, coloured and superimposed plates). It is part of a set of illustrated anatomical works dealing with human anatomy, in this case female genitalia.

It is somewhat brutal to see how the arms and legs were cut off in these anatomical illustrations as if the torso were hanging in a butcher shop. This method was apparently first used in the anatomical drawings made for William Hunter (1718–83) in the late eighteenth century. Before then the stump would have been draped in a cloth or artistically faded out. Unremitting meaty naturalism…

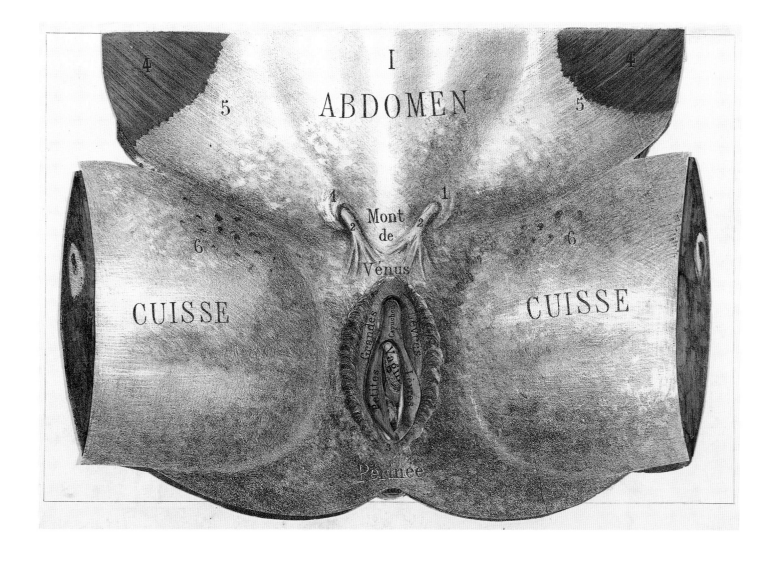

WF Gunn
*Specimens of British grasses:
with notes*
c.1893
2 vols.
Height: 46 cm (18⅛ in)
England

The joy of these two volumes lies in the way that
the grasses have been mounted on dark blue paper,
which through the contrast of colours emphasises
the beauty of the plants.

The books cover the varieties of grasses
cultivated in Britain, as well as foreign species, and
suggestions are included about how each variety
could be used for a particular purpose; for example,
as game cover, for binding sandbanks along the sea
or for the protection of wildfowl along rivers.

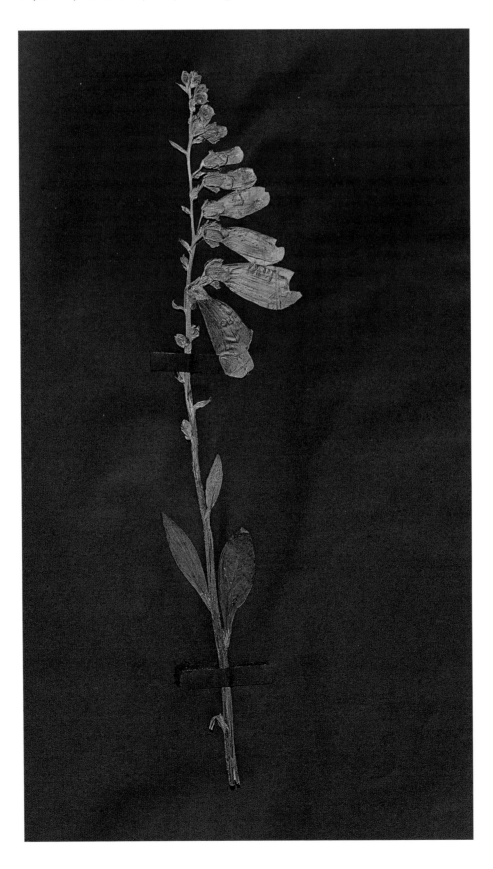

Ludovic Hirschfeld
*Névrologie ou description
et iconographie du système
nerveux et des organes des sens
de l'homme, avec leur mode de
préparation… accompagné de
quatre-vingt-douze planches
dessinées d'après nature;
ouvrage adopté par le Conseil
supérieur de l'instruction
publique*
1853
83 leaves of engravings
by JB Lévéille
Height: 28 cm (11⅛ in)
Paris

These very fine and detailed plates are another example of the kinds of teaching material that were produced in the mid-nineteenth century. As bodies for dissection were very scarce and the nervous system difficult to dissect, Ludovic Hirschfeld (1814–76) set out to provide these comprehensive hand-coloured illustrations, which deal with all the parts of the nervous system and sensory organs including how to prepare them for dissection.

Hirschfeld's work was adopted by the Conseil supérieur de l'instruction publique – the French authority responsible for textbooks in state-run institutions of learning – which meant that his work had a guaranteed market.

The illustration shown has the added charm of having the head of the dissected man included, notwithstanding his curious haircut.

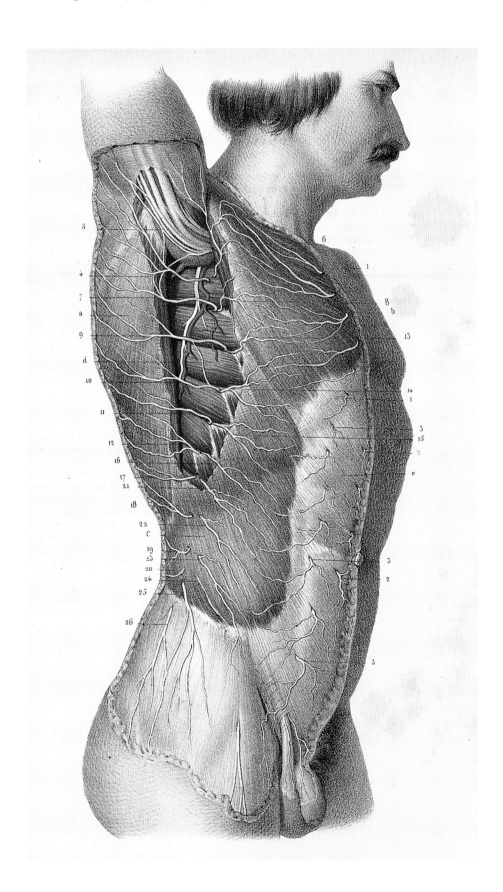

Fernand Monpillard
Four photomicrographs of
silver crystals
c.1900
Each: 21 × 15 cm (8¼ × 5⅞ in)
Paris

Fernand Monpillard (1865–1937) specialised in photomicrography and was an early researcher into colour photography.

Silver crystals are produced by running a current in a solution of silver nitrate. The size of the resulting crystals depends on the voltage and on the concentration of silver nitrate used and at this magnification the crystals end up looking rather more like plants.

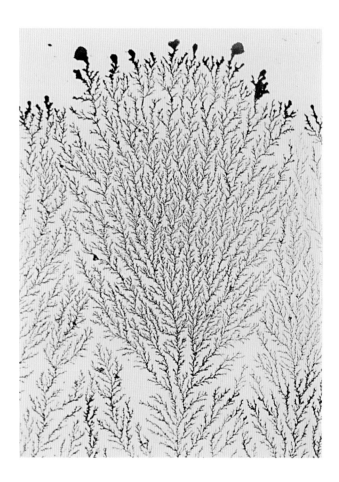
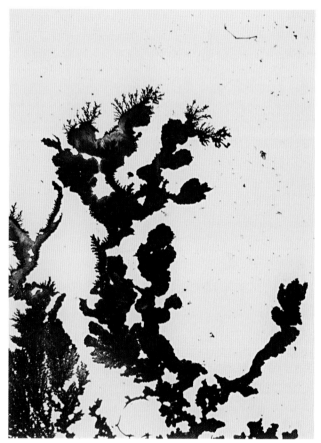

George Clarke Simpson
British Antarctic Expedition
1910–23: meteorology
1919–23
3 vols.
Height: 31 cm (12¼ in)
Calcutta

George Simpson (1878–1965) was Robert Falcon
Scott's (1868–1912) meteorologist on the fatal
1910–13 Antarctic Expedition. He was recalled to
India before Scott set out to the South Pole and
completed the meteorological volumes of the
expedition while in India. They were published
in Calcutta in 1919.

These three volumes, one of text and two of
tables and weather maps, look dull and can appear
meaningless to non-specialists, but what significance hides in these data.

Scott's team, and Simpson in particular, had
researched the weather likely to be experienced on
the way to the Pole as meticulously as possible.
Though Roald Amundsen (1872–1928), who set out
on a slightly different route, experienced no particularly unusual weather Scott was confronted
with constant storms and temperatures some ten
degrees lower than anticipated. This and the extra
man that Scott had included in the Polar Party
accounted for their food and fuel problems and
ultimately for their deaths in March 1912.

There is some suggestion that the extreme cold
in the Antarctic that year had something to do
with the extended iceberg season in the Northern
Hemisphere, when on 14 April 1912 the *Titanic*
collided with an iceberg at an unusually southern
latitude.

All this information is hidden in these tables and
if you know the story then these three volumes
exude their own form of magic. They are amongst
my favourites in the collection.

Having read quite a lot about Antarctic explora-
tion, these three volumes fairly shouted out at me
when I went to a book fair at the Royal Geographic
Society in London.

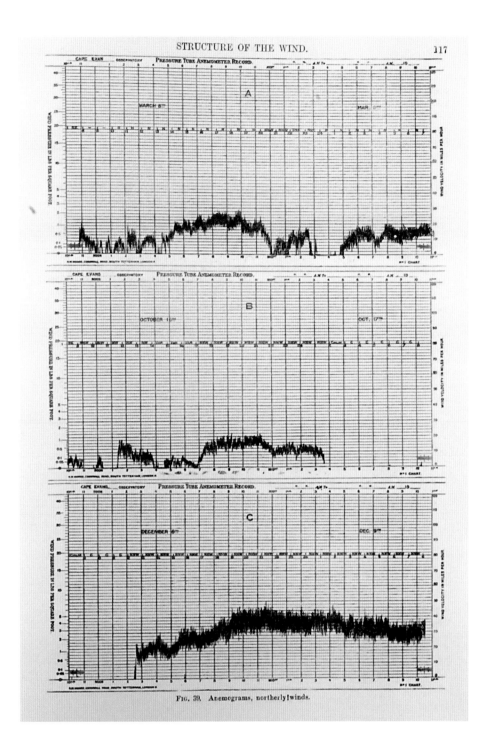

FIG. 39. Anemograms, northerly winds.

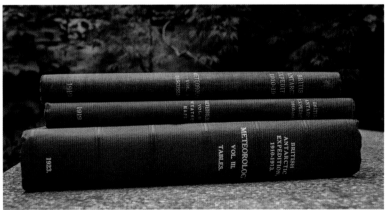

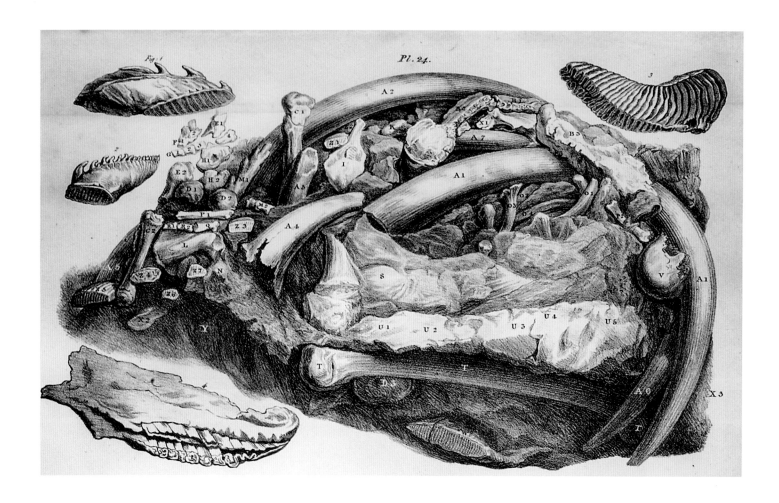

Rev William Buckland
Reliquiæ diluvianæ: or,
Observations on the organic
remains contained in caves,
fissures, and diluvial gravel, and
on other geological phenomena,
attesting the action of an
universal deluge
1823
27 leaves of engravings and
lithographs by W Buckland,
T Webster, W Clift, M
Morland, F Duncombe, H
O'Neil and G Scharf
Height: 28 cm (11⅛ in)
London

William Buckland (1784–1856) was an early
palaeontologist who had studied at Corpus Christi
College at Oxford University and was ordained
into the Church of England in 1808. Buckland
became Reader in Mineralogy at the university but
introduced increasing amounts of geology and
palaeontology into his very popular lectures. He
travelled extensively on the continent and became
friendly with Georges Cuvier in Paris, whose
opinions he quotes in this work.

Reliquiae was his first major book dealing with
animal remains from before the Great Flood,
illustrated by Buckland himself and by a number of
very competent artists. Thomas Webster (1772–1844)
was an artist, geologist and architect who studied
at the Royal Academy of Arts, London, in the 1790s.

He joined the newly founded Royal Institution in
1799 to supervise the design of the lecture theatre,
chemistry laboratory and various aspects of the
building and later took up painting and illustrating.
Mary Morland (1797–1857) was the daughter of a
solicitor who was encouraged in her scientific
pursuits by family friends in Oxford. Morland was
an accomplished illustrator who was known to
send drawings to Cuvier and others. She met and
subsequently married Buckland in 1825. George
Scharf was probably the best artist of the group
(see p.141). The originals of the beautiful drawings
he made for Buckland can still be seen in the
storage of the Oxford University Museum.
Less is known of the other illustrators: W Clift,
F Duncombe and H O'Neil.

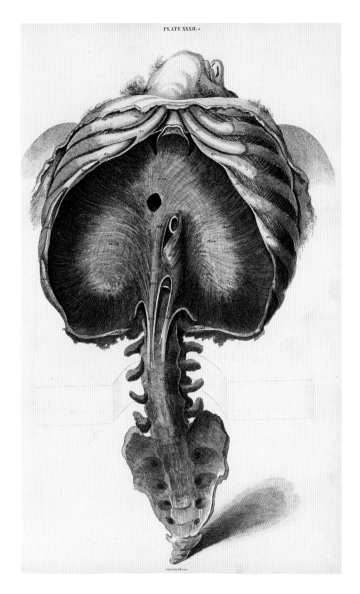

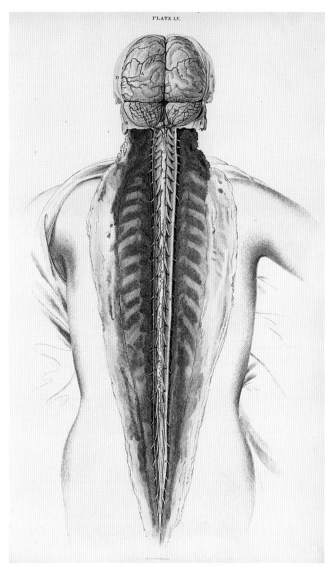

John Lizars
A system of anatomical plates of the human body: accompanied with descriptions, and physiological, pathological and surgical observations
c.1840
101 leaves of coloured engravings by WH Lizars
Height: 45 cm (17¾ in)
Edinburgh

This book dates from soon after 1832 when the Anatomy Act was passed, an act that legislated for legal access to unclaimed corpses for dissection. Prior to that only executed murderers were legally available to medical students.

John Lizars (1787–1860) was the son of an engraver, and the brother of another. After his education in Edinburgh and active service in the Peninsular War, he became a successful surgeon and teacher of anatomy in Edinburgh. In the late 1820s he is known to have lectured Charles Darwin, who was squeamish about dissecting bodies and gave up medicine.

Lizars's book of anatomical plates was an important supplement to the dissection of cadavers. The illustrations by his brother William Home Lizars (1788–1859) were intended to be as close as possible to the real thing.

The illustrations in the volume are exceptionally fine. The date that the book was purchased, 1840, is still visible. The owner of the book, whose name has been rubbed out, made copious annotations in pencil and was likely to have been a very diligent student.

Rev William Buckland
Geology and mineralogy
considered with reference to
natural theology
1836
86 lithographs
Height: 23 cm (9⅛ in)
London

This is the last work in the collection and an important one. Buckland is a great figure in early to mid-nineteenth century science, and this book reflects the balance that early scientists had to make between their discoveries and religious orthodoxy.

This book, Buckland's contribution to the *Bridgewater Treatises*, is his second in the collection (see p.248). The *Bridgewater Treatises* were a set of scientific publications intended to demonstrate the 'Power, Wisdom and Goodness of God'. They were sponsored by the Earl of Bridgewater on his deathbed and managed by The Royal Society, 'assisted' by the Archbishop of Canterbury. Eight treatises were published between 1833–40:

Thomas Chalmers, DD, *The Adaptation of External Nature to the Moral and Intellectual Condition of Man*
John Kidd, MD, *On The Adaptation of External Nature to the Physical Condition of Man*
William Whewell, DD, *Astronomy and General Physics considered with reference to Natural Theology*
Sir Charles Bell, *The Hand, its Mechanism and Vital Endowments as evincing Design*
Peter Mark Roget, *Animal and Vegetable Physiology considered with reference to Natural Theology*
William Buckland, DD, *Geology and Mineralogy considered with reference to Natural Theology*
William Kirby, *On the History, Habits and Instincts of Animals*
William Prout, MD, *Chemistry, Meteorology, and the Function of Digestion, considered with reference to Natural Theology*

The authors were all famous early scientists who to varying degrees believed what was said in the Bible.

Buckland's *Geology and mineralogy* was in a sense a compilation of much of his earlier material, including his *Reliquiae Diluvianae*. The illustrations in this work are by various artists whose work Buckland used in earlier publications. George Scharf is again prominent amongst them.

Buckland became a scientific celebrity, gaining access to influential circles in London. In 1845 he was appointed Dean of Westminster, the same year that John Henry Newman (1801–90) left the Church of England to become a Roman Catholic. The religious revivalism within the Church of England from the 1830s made it more difficult for scientists to propose ideas that differed from what was written in the Scriptures. It is said that Buckland found the going at Oxford tougher, while attendance at his lectures was dwindling.

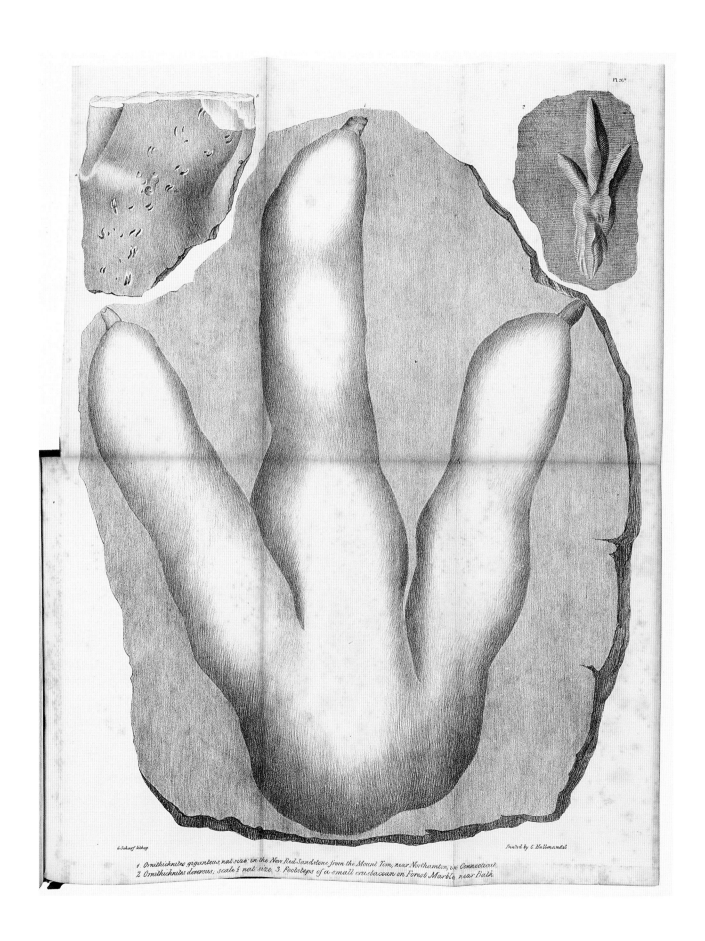

Pl.26.*

G Scharf lithog.

Printed by C Hullmandel

1. Ornithichnites giganteus, nat size: in the New Red Sandstone from the Mount Tom, near Northamton, in Connecticut.
2. Ornithichnites deversus, scale ½ nat size. 3. Footsteps of a small crustacean on Forest Marble, near Bath.

251

Acknowledgements

For development of the idea: Ken Brecher, Jonathan Drori, Errol Fuller, Michael Hoppen, Thomas van Leeuwen and Lisa O'Sullivan.

For the production: Karsten Schubert and Doro Globus, Sarah Auld, Louisa Green, Daniel Griffiths, Dorothy Feaver, Dimitrios Fragkos, Marité Lege-Germain, Alison Maurin, Chloe Nahum, Carrie Rees, James Sutton, Kitty Walsh, Steve Ward, Kate Whittington, Anna Wilson and, of course, Lynne Cooke, Robert McCracken Peck, Rosamond Purcell and Peter Willberg.

For advice, encouragement and help: David Anfam, Bergit Arends, Joanna Ebenstein, Ken Arnold, Ian Blatchford, David Breuer, Andrew Brown, Simon Chaplin, Sy Chen, Louise Franck, Michael John Gorman, Nick Hopwood, Julia Peyton-Jones, Luis Librandi, Laura Lindgren, Elsa Longhauser, Inez Lynn, Duncan McLaren, Hugh Merrell, Hans Ulrich Obrist, Vicky Paterson (and many other friends at the Natural History Museum), Henrietta Pearson, Dennis Purcell, Rebecca Rickman, Enrique Rodriguez, Ralph Rugoff, Martijn Sanders, Antony Snow, Graham Southern, MaryAnne Stevens, Dennis Stevenson, Laurens de Rooy, Kyoichi Tsuzuki, Peter Watson, Ying Yang, Masumi Yamanaka (and many other friends at Kew Gardens).

Biographies

Lynne Cooke is Senior Curator of Special Projects in Modern Art at the National Gallery of Art, Washington, DC. She was previously Curator at Dia Art Foundation, New York (1991–2008), and Deputy Director and Chief Curator at Museo Nacional Centro de Arte Reina Sofia, Madrid (2008–12). She has curated numerous exhibitions internationally and written widely on contemporary art, including artists such as Agnes Martin, Blinky Palermo, Bridget Riley, Richard Serra and Rosemarie Trockel.

George Loudon is a retired banker, collector and patron based in London. He has been a trustee of the Rijksakademie van beeldende kunsten and member of the Acquisition Committee of the Stedelijk Museum, both in Amsterdam. In 1995 he was a judge of the Turner Prize. He is currently a trustee of the Royal Botanic Gardens, Kew.

Robert McCracken Peck is Curator of Art and Artifacts and Senior Fellow at The Academy of Natural Sciences of Philadelphia, Drexel University, Philadelphia. He has served as a guest curator and consultant to museums and libraries throughout the US and has written for *Nature*, *Natural History* and the *New York Times*, among others. His books include *A Glorious Enterprise: The Academy of Natural Sciences of Philadelphia and the Making of American Science* (University of Pennsylvania Press, 2012).

Rosamond Purcell is a photographer whose work often engages with the museum and the nature of collecting. Her works have appeared in numerous publications and scholarly journals, and she collaborated frequently with the Pulitzer Prize-winner Stephen Jay Gould. Her books include *Egg & Nest* (with Linnea S Hall and René Corado, Harvard University Press, 2008) and *Dice: Deception, Fate & Rotten Luck* (with Ricky Jay, WW Norton & Company, 2003).

Published in 2015 by **Ridinghouse**

46 Lexington Street
London W1F 0LP
United Kingdom
ridinghouse.co.uk

Ridinghouse Publisher: Doro Globus
Publishing Manager: Louisa Green
Publishing Assistant: Daniel Griffiths

Distributed in the UK and Europe by
Cornerhouse Publications
c/o Home
Manchester M1 5NH
United Kingdom
cornerhousepublications.org

Distributed in the US by
RAM Publications + Distribution, Inc.
2525 Michigan Avenue Building A2
Santa Monica, CA, 90404
United States
rampub.com

All photography by Rosamund Purcell unless
noted below:
Steve Ward: pp.2, 10–11, 13, 28–31, 35, 43, 93, 94, 96,
    97 (left), 99–100, 102, 110, 114 (right), 115, 119 (left),
    122 (top), 126–27, 131–32, 136, 137 (left), 138, 140, 142,
    154, 158–59, 161, 172–74, 180, 182, 185, 198, 203, 206,
    214 (top), 221, 228–29, 232, 241–46, 249, 251
Efstratia Verveniotou and Liesa Brierley: pp.210–11

British Library Cataloguing-in-Publication Data:
A full catalogue record of this book is available
from the British Library

ISBN 978 1 909932 10 4

Designed by Peter Willberg with James Sutton
Set in Fedra Sans
Printed in Italy by Castelli Bolis

Project managed by Sarah Auld
Assistant to George Loudon: Alison Maurin
Proofread by English Editions

Cover:
Reconstituted *Aepyornis maximus* (Elephant bird) egg
(see p.88)

Editorial note:
Not all items in the collection are illustrated.
Information that is not known is not listed. In some
cases, titles and author listings correspond directly
to the information stated on the objects. Titles of
printed matter are listed in their original language.
Height precedes width precedes depth. In some
cases, height precedes diameter only. Entries with
multiple objects or parts are listed with dimension
ranges. In line with the conventions of book
collecting, only height is listed for publications.

Ridinghouse